THE · PERFECT · MACHINE

THE · PERFECT · MACHINE
TV IN THE NUCLEAR AGE

JOYCE NELSON

Author's Note:
Portions of this work first appeared, in
slightly different form, in *This
Magazine, Cinema Canada,
Canadian Drama, The Globe and Mail,
Borderlines,* and *The New
Internationalist.*

Published by Between The Lines
229 College Street
Toronto, Ontario M5T 1R4

Cover illustration 'Please change the channels' by Richard Slye
Cover design by Goodness Graphics
Typeset by Coach House Press, Toronto
Printed in Canada

Between The Lines
receives financial assistance from
the Canada Council, the Ontario Arts Council,
and the Department of Communications.

Canadian Cataloguing in Publication Data

Nelson, Joyce, 1945-
    The perfect machine

Includes bibliographical references.
ISBN 0-919946-84-4 (bound).
ISBN 0-919946-85-2 (pbk.)

1. Television – Social aspects.
2. Television – Psychological aspects.
3. Nuclear industry. I. Title.

PN1992.6.N45 1987   302.2'345   C87-094598-X

Our generation is a bridge generation attempting to make a giant stride in consciousness. Faced with atomic power, faced with the possibility of our own self-destruction, we are trying to reconnect to roots that have lain dormant underground for centuries in the hope that the nourishment from those depths may somehow counterbalance the sterility of the perfect machine.

MARION WOODMAN, *Addiction To Perfection*

Contents

Acknowledgements

MAY I EXPRESS my thanks to a number of individuals who helped immensely throughout the writing and compiling of this text: Robert Clarke, who edited the manuscript and believed in its worth as a project; the Between The Lines collective, who made its publication possible; the "Thursday Group" and the *This Magazine* editorial collective, both of which provided various kinds of support for the writing; Marlene Bird and Vicky Kosharewich, who provided technical assistance as well as many other favours; Naomi Bennett, John Went, Connie Tadros, Michael Dorland, Geraldine Sherman, all of whom have been encouraging in their own ways; and the people of Ontario, without whose financial assistance through the Ontario Arts Council the manuscript could not have been completed.

J.N., *May 1987*

1

Introduction: The Bolt in the Soul

It is my judgement in these things that when you see something that is
technically sweet, you go ahead and do it and you argue about what
to do about it only after you have had your technical success. That is
the way it was with the atomic bomb.

J. ROBERT OPPENHEIMER

U.S. TELEVISION is now prepared for World War III. The Federal Emergency Management Agency (FEMA), which has the responsibility of establishing contingency plans to ensure that institutions like the U.S. government (down to the state and local level), the IRS, and the Post Office will
continue efficiently after a nuclear holocaust, has issued a made-for-TV
film to be shown in the event of an international nuclear crisis. This
twenty-five-minute animated civil-defence film, *Protection In The Nuclear
Age,* has been distributed across the United States to local civil-defence
officials, many of whom have dutifully passed the copies on to their local
television stations. Since this is probably the last telecast that most Americans, and many Canadians (70 per cent of whom are now tuned to U.S.
channels through cable-TV) will see, the film attempts an upbeat tone.

Produced in English and Spanish versions, with captions for the deaf,
the film opens with a drawing of planet earth as the voice-over calmly
announces: "We live in a world of tension and conflict. And peace, even
where it does exist, does so without guarantees for tomorrow." The

camera zooms in on the North American continent as the voice-over continues: "We must therefore face the hard reality that someday a nuclear attack against the United States might occur. And – equally important – we must also realize that horrifying as that prospect may seem, destructive as such an attack might be, we can survive. It would not mean the end of the world, the end of our nation. And you can greatly improve your own chances of survival if you'll remember these facts – about Protection in the Nuclear Age."[1]

It should not surprise us to learn that right up through the last telecast, television is prepared to convey optimism about nuclear Armageddon. That has been its role throughout its history – a history that coincides with the development of a nuclear industry bent on planetary destruction. Because virtually the same corporations developed TV and the full-blown nuclear arms and nuclear power systems, the two mass media that dominate our age – television and the bomb – cannot help but be intertwined in an ideological embrace. The parameters of that embrace are larger than either medium, since both are the culmination of a mind-set succinctly expressed by J. Robert Oppenheimer, whose scientific expertise fuelled the Manhattan Project. That mind-set, patriarchy, transcends all lesser ideologies, as recent events remind us.

□

In October 1986, at the Iceland Summit, the United States effectively scuttled an opportunity to reach an agreement with the Soviet Union on the dismantling of nuclear weapons. This historic possibility apparently could not compete with the allure of "Star Wars" – the U.S. Strategic Defense Initiative which so bedazzled Reagan and his administration. For its part, the Soviet Union could, as of a few months earlier, claim responsibility for the greatest nuclear accident in history. The atmosphere and soil of the planet, as well as our own bodies, are still absorbing the effects of Chernobyl. Of course, those effects nicely combine with the resumption of underground nuclear testing by the U.S. Department of Energy: tests designed, according to Energy spokesmen, to determine the ability of U.S. space and military hardware to survive a nuclear blast.

That same Department of Energy is developing the solar-powered satellites slated to be part of the orbiting space stations heralded for the 1990s. Dr. Rosalie Bertell, an authority on the nuclear industry, states:

Sixty space stations are planned in this solar-powered satellite system (SPS), each of which is estimated to cost one trillion dollars. The SPS would work by concentrating the sun's rays, using laser and microwaves, down to a rectenna (a receiving antenna) on earth. The rectenna would convert solar energy to electricity. The SPS is also capable of wiping out all communications systems in a city, including computer memories, and can destroy ballistic missiles in air. It is an incendiary weapon and an antipersonnel weapon. The microwave can kill people and save buildings.[2]

Utterly dispensable in the path of all this technical success, people and other living things have been reduced to the status of objects, while space and military hardware and technologies have been endowed with the status of subjects: indispensable, valued, worth massive investment and protection from harm. With the world's leaders currently spending more than $45 million every half-hour on such "death-technologies" (to use Petra Kelly's term), it is hard not to recognize the terrible, harsh priorities that characterize our age.

In his comprehensive critique of modern science and engineering, *The Turning Point,* Fritjof Capra states:

> The psychological background to this nuclear madness is an overemphasis on self-assertion, control and power, excessive competition, and an obsession with "winning" – the typical traits of patriarchal cultures.... Nuclear weapons, then, are the most tragic case of people holding onto an old paradigm that has long lost its usefulness.[3]

It is far easier to gauge the results of this old patriarchal paradigm than to account for its psycho-historical origins. As the last two decades of the twentieth century grind to their end, we are surrounded by what J. Robert Oppenheimer would call the "technically sweet". A vast array of new technologies now intervenes to control life from cradle to grave and indeed – in the case of test-tube conception, genetic engineering, cryogenics, and suspended animation – from before the cradle to beyond the grave. There is no question that our scientists and engineers have gone ahead, à la Oppenheimer, and achieved their successes unhampered by considerations beyond technical sweetness.

Many of these technological successes are historically part-and-parcel with work on the atomic bomb in the early 1940s. For example, both genetic engineering and the computer revolution are direct off-shoots of the Manhattan Project, involving many of the same personnel for their research and development. (Perhaps not surprisingly, both technological developments expressed a central, ironic contradiction. While the computer was necessary to make the bomb, genetic engineering was foreseen as way of "fixing" the damage done by it.) Combined with other war-generated practices – the heavy reliance upon petrochemicals, changes in agricultural practice towards monoculturing and "supercrops", increased automation, rapid expansion in biotechnology – these new developments confront us with a modern arsenal of technique that staggers the mind.

Behind these technological successes it is possible to detect an all-out war on nature itself, including human nature. They suggest a ruthless dissatisfaction with the natural world, whose "mistakes" must be corrected in the name of progress. As Glenn Seaborg, former chairman of the U.S. Atomic Energy Commission, once explained: "Large nuclear explosives give us, for the first time, the capability to remedy nature's oversights."[4] Thus, a river flowing the "wrong" way according to an engineering plan may be nuked into submission and redirected to suit the needs of progress.

Of course, if one asks "Progress for whom, and under what conditions?" the picture is altered significantly. For example, Kevin Robins and Frank Webster, in their assessment of high-tech, write: "There are clear indications that microelectronics (information technology) is being constituted not as a liberating technology, but as one that will facilitate the rule of capital across ever wider spheres of social existence."[5] But to question technological advance and expansion is to risk being labelled anti-progress.

Obviously, the word "progress" itself masks many things, including an article of faith that is central to the old patriarchal paradigm. That article of faith has been summarized by Jeremy Rifkin in his book *Declaration Of A Heretic*:

> For the first time in recorded history, a new value has emerged, and in a very short period of time it has effectively eclipsed every other deeply held traditional value that the human species has entertained over the ages. The value is efficiency ... the one universally held value that cuts across all ideologies and geographic boundaries.[6]

Efficiency is a technological value, a machine characteristic. Summarized in the word are the strictly mechanical goals of speed, perfection, and quantitative results. Whatever can achieve these goals better than a previous means is considered to be a sign of progress, regardless of its effects upon other aspects of living.

As a concept and value, efficiency seems unlimited, for there are always potential techniques that may be discovered to improve efficiency. Like progress, of which it is the sign, efficiency is always incomplete: luring its advocates on to greater effort. Efficiency is thereby the watchword of engineering – that field of endeavour that is quite literally the engine of modernity. As Rifkin notes, "Engineering is a process of continual improvement in the performance of a machine, and the idea of setting arbitrary limits to how much 'improvement' is acceptable is alien to the entire engineering conception."[7]

With the value of efficiency eclipsing all others, we have come to live according to its dictates: engineering ourselves and all life towards greater and greater production, perfection, and performance. It is in this sense that technology is not "neutral" but instead permeates the very tenor of life. George Grant states it well:

> When we represent technology to ourselves as an array of neutral instruments, invented by human beings and under human control, we are expressing a kind of common sense, but it is a common sense from within the very technology we are attempting to represent.... We are led to forget that modern destiny permeates our representations of the world and ourselves. The coming to be of technology has required changes in what we think is good, in what we think good is, in how we conceive sanity and madness, justice and injustice, rationality and irrationality, beauty and ugliness.[8]

The value of efficiency – the supreme value in a technological age – prevents us from fully recognizing the toll upon nature, our bodies, our feelings, our souls. Indeed, these dimensions come to be regarded from within the same efficient mind-set: as utterly inferior, "irrational" aspects that need to be whipped into shape and harnessed for better efficiency, or else denied entirely until illness, malaise, stress, or environmental crises remind us of our human nature within a natural ecosystem.

Standing in the way of efficiency, standing in the way of this sign of progress, is human corporeality: with all its humbling imperfections, its slow rhythms, its tie to the natural order. This clumsy impediment to progress, this embarrassment to patriarchal efficiency, is the despised object that lies across the path of modernity, blocking its smooth progression. "Disinherited and separated, the body," writes British literary historian Francis Barker, "is traduced as a rootless thing of madness and scandal and then finally, in its object-aspect, it is pressed into service."[9] But the body resists this servitude, this objectification. The despised body returns from its exile to assert another order of values beyond efficiency. Jungian analyst Marion Woodman states:

> We are living in a technological age that puts its faith in the perfection of the computer. Human beings tend to become like the god they worship, but fortunately for us, our agony does not allow us to become perfect robots. However hard we try to eradicate nature it eventually exerts its own value system and its own painful price.[10]

This human factor (and it has been reduced to a "factor") has always been problematic for efficiency experts in all walks of life, and not surprisingly we are now witnessing the rapid elimination of human beings from more and more dimensions of existence. As Ursula Franklin observes: "Contemporary technological societies are establishing an anti-people climate where people are increasingly looked upon as sources of problems and technological devices as sources of solutions."[11]

Within the patriarchal paradigm we must include this underlying contempt for the natural human body in all its "inefficient" imperfection. As the site of grounding wisdom, the body continually reminds us of our limitations. But limitations are precisely what progress and technological efficiency defy. Now that we have reached a "height" of technological progress and efficiency that has brought us to the brink of planetary destruction, we may well ask what is behind this patriarchal defiance of limitations. What fuels this underlying contempt for corporeality and nature itself?

Fortunately, there is a text familiar to most of us which deals with such questions, exploring in archetypal terms the deeper dimensions of the patriarchal *zeitgeist*. Its author – the daughter of political philosopher William Godwin and feminist author Mary Wollstonecraft – has chillingly articulated the paradigm that structures modern times.

□

Mary Wollstonecraft Godwin Shelley's extraordinary first novel, *Frankenstein*, published in 1818, has at its centre a vivid portrait of the modern scientist / engineer obsessed with the "technically sweet". Her Victor (an ironic pun) is initially a young, self-taught student intrigued by the writings of the alchemists of the Middle Ages. But when he leaves his village to enter university, Victor Frankenstein quickly learns that his personal study is considered woefully outdated: the cosmology of the alchemists has been overthrown by the Scientific Revolution. His professor of modern chemistry, M. Waldman, especially impresses him with a panegyric on the wonders of modern science compared to the "failures" of the alchemists:

> "The ancient teachers of this science," said [Waldman], "promised impossibilities and *performed nothing*. The modern masters promise very little; they know that metals cannot be transmuted, and that the elixir of life is a chimera. But these philosophers ... have indeed *performed miracles*. They penetrate into the recesses of nature, and show how she works in her hiding places.... They have acquired new and almost unlimited powers; they can command the thunders of heaven, mimic the earthquake, and even mock the invisible world with its own shadow."[12]

Carl Jung and Marie-Louise Von Franz have explored the work of the alchemists as a profound symbolic grappling with three key dimensions:

the tension of opposites, the role of the feminine, and the sacredness of matter. As a practice, alchemy appears to have been devoted to an inner transformation moreso than an outer one. Through their experiments and ritual, the alchemists were uncovering inner truths (that had outer correspondences), truths that formed a paradigm in which the earthly matter, the feminine, and the balance of opposites were central to finding the "gold" within living itself. As a world-view, alchemy was profoundly balanced and harmonious.

Mary Shelley seems to have intuited this understanding of alchemy; she has her young hero, Victor, begin his personal study within the science only to have this world-view overthrown. Impressed with the tenet that it is outer performance that counts in modern science – indeed, the performing of miracles and the acquiring of "almost unlimited powers" – Frankenstein is ripe for the project that will soon possess him. But before her creation is seized by the technically sweet feasibility of "infusing life into an inanimate body", Mary Shelley is careful to subtly reveal for us the degree to which there are no boundaries, either in Victor's scientific milieu or within himself, to limit his endeavours.

Waldman, his esteemed new mentor, seemingly gives him experimental carte blanche by assuring him, "The Labours of men of genius, *however erroneously directed,* scarcely ever fail in ultimately turning to the solid advantage of mankind."[13] Moreover, it is clear from Waldman's speeches that modern science suffers no restrictions in its drive to "penetrate" nature's secrets: "commanding", "mimicking", "mocking" the natural world. Shelley's text, it seems, is primarily concerned with patriarchal hubris: the tragic flaw in modern consciousness itself.

Similarly, Frankenstein admits to having no personal sense of awe or respect for mysteries beyond human understanding, beyond the rational intellect. As he says, "In my education my father had taken the greatest precautions that my mind should be impressed with no supernatural horrors. I do not ever remember to have trembled at a tale of superstition, or to have feared the apparition of a spirit. Darkness had no effect upon my fancy."[14]

In establishing this lack of boundaries, whether within modern science or within Victor himself, Shelley seems to be pointing to the hallmarks of the Scientific Revolution: the supremacy of the completely rational intellect untrammelled by any sense of the sacred or by any respect for the so-called "irrational" dimensions of nature or the human being. Thus, in contrast to the world-view of alchemy, modern science excludes any sense of sacredness in matter, excludes any significant role for the feminine principle in life, and refuses the balance of opposites necessary to existence. Cut off consciously from these tempering forces, Frankenstein is thereby completely subject to the psychic inflation that overtakes him. He is compelled to break that most ancient boundary delimiting humanity from the Deity: "Life and death appeared to me ideal bounds, which I should first break through, and pour a torrent of light into our dark world."[15]

A major structuring principle in the novel is this opposition between light and dark, which Shelley uses for thematic and ironic purpose. Darkness is clearly associated with the ancient mysteries and hidden recesses of nature which modern science and the supremely rational consciousness are so intent upon penetrating and exposing. But especially, darkness is associated with nature's greatest mystery: death. It is this darkest mystery that Frankenstein (and the penetrating light of science) seeks to banish.

> I paused, examining and analysing all the minutiae of causation, as exemplified in the change from life to death, and death to life, until from the midst of this darkness a sudden light broke in upon me – a light so brilliant and wondrous, yet so simple, that while I became dizzy with the immensity of the prospect which it illustrated, I was surprised, that among so many men of genius who had directed their enquiries towards the same science, that I alone should be reserved to discover so astonishing a secret.[16]

Thoroughly dazzled by this enlightenment, as well as by his own genius, Frankenstein must pursue it beyond the realm of insight into that active arena of modern science: the performing of miracles. What can be done, must be done. And while he admits to no supernatural horrors in the face of great mysteries, he readily confesses "an almost supernatural enthusiasm" for the performance that must follow the discovery of "so astonishing a secret".[17]

It is precisely here that Mary Shelley's own particular genius reveals itself. Rather than focus her narrative upon delineating the gradual breakthroughs by which Frankenstein achieves his desired technical success, she concentrates instead upon detailing the gradual breakdown and deterioration of the human Victor. What Shelley shows us is an intellect, a consciousness, increasingly divorcing itself from the realm of the living. Following the mad light of his supreme rationality, Frankenstein gradually becomes completely estranged from all human society, embodied in his family, his friends, and his university peers. Even more tellingly, he abandons his intimate connection to the woman he loves: Elizabeth. Living totally isolated in his "cell", Frankenstein's only forays into the outside world are for the purpose of securing corpses for his experiments. His attempts to break the bonds of life and death place him in the realm of death itself. Within the dead of night he forages for corpses to bring back to his laboratory.

Though he is at times dimly aware that his monomaniacal goal is some-how robbing him of all that he had once held dear, Victor cannot stop him-self. He is a man possessed.

I could not tear my thoughts from my employment, loathsome in itself, but which had taken an irresistible hold of my imagination. I wished, as it were, to procrastinate all that related to my feelings of affection until the great object, which swallowed up every habit of my nature, should be com-pleted.[18]

Hypnotized by the sheer technical sweetness of an otherwise loathsome project, Frankenstein can neither consider its consequences nor fully per-ceive the toll it exacts from his own health. Unable to rest or eat, vaguely aware of a terrible nervousness overtaking his bodily self, he forges on, despite the irony that his goal of infusing life into an inanimate body is simultaneously draining the life from his own. And what is the perceived pay-off to Frankenstein for this living death?

A new species would bless me as its creator and source; many happy and excellent natures would owe their being to me. No father could claim the gratitude of his child so completely as I should deserve theirs. Pursuing these reflections, I thought, that if I could bestow animation upon lifeless matter, I might in [the] process of time ... renew life where death had apparently devoted the body to corruption.[19]

Obsessed with a gratitude owed to the deserving Father, fascinated with birthing his own unnatural simulacra beyond the reach of death, Victor Frankenstein eventually destroys, in the process, everything that had once made his life meaningful. "The bolt has entered my soul," he mutters, voic-ing the condition that characterizes patriarchal modernity.[20]

In Jungian terms, we could say that Frankenstein has totally ostracized the feminine side of the human psyche: the part in us all that is deeply rooted in human feelings, bodily instincts, the rhythms of nature, and a sense of containment within the larger forces of earthly life. As a result of this splitting-off of the feminine in himself, Frankenstein's masculine ener-gies – which excel at rational thinking, goal-setting, and performance – soar off unbounded, wreaking havoc on the bodily self and the natural world. This psychic split mirrors, and results from, the mind-body split. Like Descartes, upon whom he may have been modelled, Frankenstein could well state: "There is nothing included in the concept of body that belongs to the mind; and nothing in that of mind that belongs to the body."[21] The attempt to defeat death (a condition abhorrent to the mind) is an attempt to defy and defeat the body. Ironically, that very attempt brings death. By trying to fashion a new species of simulacra beyond death's reach, Frankenstein makes of his own life a living death and creates the means by which his loved ones are destroyed.

Death – the dark and terrible mystery, the final limitation on personal and societal hubris – was, in non-patriarchal cultures, fully respected and even honoured as a part of the whole life process. Before this deep mystery, the mind must yield to a transcendent power paradoxically immanent in the natural world and the human body itself. This grounding wisdom of the body-mind was, in past ages, the fully humbling and ennobling insight (or knowing within) that situated humanity inside and within nature, not apart from it.

But patriarchy refuses to be thus situated, refuses this death which makes us siblings to all other living species. As Ernest Becker writes:

[Man's] body is a material fleshy casing that is alien to him in many ways – the strangest and most repugnant way being that it aches and bleeds and will decay and die. Man is literally split in two: he has an awareness of his own splendid uniqueness in that he sticks out of nature with a towering majesty, and yet he goes back into the ground a few feet in order blindly and dumbly to rot and disappear forever.[22]

The existential pain in this passage seems unique to modernity. A past, non-patriarchal age would not have conceived of mankind as in any way "sticking out" of nature with a "towering majesty" – not because such past cultures were in any way less complex or intelligent than ours, but because such societies perceived the human species as clearly different from, but not superior to, other species. Granting every living thing and the earth itself a "splendid uniqueness", non-patriarchal cultures did not conceive of "conquering nature" or of "commanding" or "mocking" it. But, as Elizabeth Dodson Gray has revealed in her landmark book *Green Paradise Lost,* where there is actually only difference, patriarchy perceives hierarchy.[23]

Loathing the fate of the body, the patriarchal mind must usurp the greatest "uniqueness" for itself: ranking all living things according to a hierarchy that places humanity (and especially the white male for his alleged rational prowess) just below the Diety and fully master of all the rest of creation. It is precisely here, in this terrible desperation for greater uniqueness, that the patriarchal mind splits off from nature, splits off from the body: making of it an alien, fleshy casing, strange and repugnant.

With the "death of God" – that sense of the sacred which once held this "great chain of being" in place and thus somewhat circumscribed the endeavours of science and engineering – the patriarchal mind moves into its full desired stature: unlimited, unboundaried, remaking the world into an "anti-people" laboratory filled with its own simulacra which parallel the "new species" dreamed of by Frankenstein: blessing their human creator, grateful to the deserving Father, happy, efficient, excellent, and beyond death.

□

In his book *Technology and the Canadian Mind* Arthur Kroker concludes, "In ways more pervasive than we may suspect, technology is now the deepest language of politics, economy, advertising, and desire." He argues:

> It is apparent, now more than ever, that we are living in the midst of a terrible *ethics gap*: a radical breach between the realities of the designed environments of the new technologies, and the often outmoded possibilities of our private and public moralities for taking measure of the adequacy of technological change. It's as if we live in a culture with a super-stimulated technical consciousness, but a hyper-atrophied moral sense. It is just this gap between ethics and technology which makes it so difficult to render meaningful judgments on specific technological innovations in satisfying or thwarting the highest social ideals of western culture.[24]

In disseminating this "super-stimulated technical consciousness", television plays a primary role. We tend to think of the medium as a "window on the world", without recognizing that it is more obviously a technological cataract. As Kroker puts it, "It is a distinctively modern fate to *live* technology as a kind of second biology."[25] Given the underlying hatred of the first biology (nature and the body) under patriarchy, and given the overwhelming pervasiveness of electronic media, it seems more accurate to say that it is our modern fate to live technology.

If television is, on the one hand, a technological cataract remaking our vision and sensorium in subtle ways, it is also a probe into our unconscious.

We come after Freud and Jung, so we know that the deepest layers of our psyche communicate to our conscious selves through a language of images spontaneously arising in dreams and waking fantasies. The numinous power of such images arising unbidden and involuntary to consciousness has long been recognized not just by twentieth-century psychiatrists, but by artists and poets, shamans, priests, and visionaries throughout

history. The rituals, vision quests, and creative cultural expression of earlier tribal societies were primarily based on the recognized power of spontaneous or ritualistically-evoked images rising up from the unconscious.

Similarly, past cultures seem to have been fully aware that the *path to*
the deeper layers of the psyche is through humanly-created images. The
recognition of this powerful pathway is apparent in Old Testament prohi-
bitions against "graven images", but it is also found, in other cultures and
other eras, in the selection of certain proscribed images for contemplation:
mandalas for meditation, or sacred pictures to inspire reverential prayer,
for instance. Even those earliest cave drawings of animals sketched in red
ochre may have served as numinous images to deeply attune members of
the tribe to the hunt. In past ages, the humanly-created image seems to
have been fully intuited as a highly powerful and affective pathway or trigger to deeper realms of the psyche beyond conscious rational thought.
That intuition placed an aura of sacredness around the image and set
boundaries around its use in the culture.

By contrast, we now live in an extremely paradoxical age with regard to
this power of images. On the one hand, our very rational approach to life
dissuades us from personally developing any connection or relation to our
own unconscious. We are taught to dismiss the dream images that arise to
consciousness each night, as well as our spontaneous waking fantasies, as
so much "irrational" nonsense. Under patriarchy, the conscious rational
intellect is valued over all else: over feelings, intuitions, bodily wisdom,
but also over the unconscious itself, that dark realm feared and despised by
the strictly rational mind.

But, on the other hand, we live in an age that completely immerses us in
a great flood of images, many of which have been purposely designed to
exploit the unconscious. In the midst of this great sea of images – the
stock-in-trade of network TV, cable TV, pay-TV, satellite-TV, MTV, incessant
advertising, billboards, movies, video-cassettes – we seem to have forgotten what even the most "primitive" culture once knew and respected: that
the image is a pathway to our deepest psychic levels and is thereby an
incredibly powerful thing to be used wisely, even ritualistically, because of
its potent resonance with the unconscious. Of course, such cultures used
other terms – the underworld, the realm of the gods – but were referring to
that same realm of immense energies that we now term the unconscious.

This great paradox places us all in an extremely vulnerable position.
The irony of our patriarchal age is that at the same time as we have been
taught to discount what the patriarchy scorns as "irrational", a great
industry has risen up to fully exploit those same scorned dimensions. It is
precisely here that patriarchy and capitalism meet: at the site of exploiting
what is scorned. Television is the epitome of that exploitation industry:

especially because of its location in the home and its non-stop, twenty-four-hour a day accessibility.

It is entirely possible that the very fact of this immense flood of images all around is itself a sign that our age hungers for reconnection to the deeper layers of the psyche. The long reign of patriarchy, with its hyper-rationality and scientific stance, has steadily severed our roots in unconscious depths. But we simply are not nourished by the rational alone. As Edward C. Whitmont writes in *The Symbolic Quest*:

> This activity of consciousness – the establishment of control in the world of things through conceptualization, rational thought, and the development of discipline and the abstractive repression of emotions – is an utterly vital and indispensable phase of psychic development. It leads from psychic primitive infancy to adulthood. Mythological tradition likens this development to the creation of the world from the original chaos to the establishment of a foot-hold on dry land away from the threat of drowning in the flood waters. Yet it is not the "dry land" of rational consciousness that contains and supports the ocean but conversely: the waters of the ocean contain the dry continents, and life upon them depends upon the waters. Similarly, it is the … unconscious psyche that gives rise to and maintains the world of consciousness. Consciousness with its concepts is a relatively minor part of the total psychic functioning, and in terms of dynamics certainly not the most powerful one. It establishes fixed points of rational reference – but at the price of a loss of emotional connection. *Images*, on the other hand, constellate emotional and imaginative qualities and thus *reconstitute a connection* which the abstractive process has severed.[26]

Thus, the image-making technologies that have been invented over the past century – photography, film, television – may themselves be a sign of an age trying to reconnect, via the mass-produced image, to the unconscious.

But there are inherent problems in this technological reconnection. Immediately apparent is that these technologies constitute us as spectators in relation to both the image and the world: a site of alienation ideally suited to our own exploitation by the capitalist industries that cohere around the technologies. As a mass audience of spectators, our emotional and imaginative reconnection to the unconscious is affected by powerful interests which have their own ideological goals in mind.

The danger in our time is that the great flood of contrived images all around us purposely serves to sweep the individual psyche back into a kind of "psychic primitive infancy" where the conscious stance is undermined by the manipulative releasing of unconscious flood waters. For children and others still emerging into individual consciousness, this great omni-present sea of images must surely at some level impede the development of those necessary "fixed points of rational reference", those "footholds on dry land" so crucial to psychic health. But none of us is invulnerable. It would be more insightful to say that our patriarchal capitalist society pos-its things themselves – consumer items, technologies, and globalized tech-nological systems – as the "fixed points", the "footholds" by which the individual finds momentary stability.

At the level of product consumption, it is obvious that capitalism depends on built-in obsolescence and the unceasing replacement of the new, so that those apparent fixed points and footholds are forever chang-ing. The individual is intended to be always scrambling, so to speak, for the next piece of "dry land". But at the level of technological systems – and here we must include not just the computer "revolution" and the media "explosion", but also the immense technological advance evidenced in nuclear weaponry – such systems are themselves posited as the greater "fixed points" of society, the enduring "footholds" for the human collec-tive in the postmodern age. Here we begin to perceive the "faith" in sys-tems like "Star Wars", the "trust" in nuclear technologies, the "belief" in televised images.

This should alert us to an even greater vulnerability of our time. As societies we have lost what Jungians refer to as the "ritual container" nec-essary for social and psychological health. In past ages, given collectives of people had their own unique face-to-face rituals by which each member felt contained and valued within the social matrix. Fully participatory and communal rituals – like story-telling, song, dance, religious festivities, apprenticeship in a craft, rites of passage to mark the stages in life – all pro-vided the shared "container" for the immense conscious and unconscious energies of a people. The traditional rituals channelled these energies along pathways that all met at a shared centre of meaning, purpose, and har-mony. Individuals felt themselves to be contained, body and spirit, within the larger matrix: encompassed in meaning, and contributing to that meaningful whole through the living out of their individual lives.

While earlier societies provided this "ritual containment" for their people, the growth of patriarchal capitalism has almost completely elim-inated not just meaningful face-to-face ritual from our lives, but meaning itself. The emphasis upon "rugged individualism" and competition and the rational exclusion of mythic and sacred dimensions of living have left the individual virtually alone, and uncontained, in a seemingly hostile and absurd world. This existential alienation, so uncharacteristic of so-called "primitive" societies, undoubtedly coincides with our alienation from the body. Uncontained in any meaningful way, we are especially vulnerable to ideological and / or technological systems that promise containment for our vulnerable selves.

□

Rapid changes through this century, escalating during World War II and in the postwar period, have altered every aspect of society, particularly those time-honoured social institutions that once provided some form of stability and containment for the individual: church, family, community, meaningful work, pageantry, the arts. In many ways, it is to our benefit that such institutions are crumbling under the impetus of change. For many people, the church and the nuclear family (for instance) have been strait-jackets of oppression. The problem is that these social changes have been accompanied by, if not instigated through, a tremendous rise in the power and hegemony of the mass media. The spiritual and social vacuum created by the demise of traditional Western institutions has steadily been filled, not by alternative social arrangements or more viable forms of face-to-face ritual, but by the media themselves, which have rushed in to fill the breach. This is especially true of television, which for millions and millions of people is church-family-community-pageantry-the arts all rolled into one.

Television has assumed extraordinary ritual and institutional power in our lives: not just by commanding some seven hours a day of our leisure time (the U.S. average), but by becoming the new "matrix", the new "ritual container" for most people. Culture has become almost entirely television-culture, or television-derived "culture"; ritual is now television "ritual": we follow its change of seasons, the rhythms of its broadcast "day", we attend to its pageantry and festivals, its stories, songs, and dance – all of which have been created by a few corporate conglomerates in an industry fully aware of its institutional power, and the power of the screen.

As most people outside the United States know, that industry power extends far beyond U.S. borders. What may have begun as a technological "ritual container" for American society in the aftermath of World War II has, through specific industry practices and business structure, been extended across the globe. If patriarchy and capitalism meet at the site of exploiting the body and scorned dimensions of the human psyche, we might also say that patriarchal capitalism and imperialism meet at the site of the screen. The term "global village" has been terribly misleading. It prevents us from fully recognizing that one country, the United States, dominates that "village" and does so for ideological goals that are beyond simple profit. In this century colonization is accomplished through the eye. At least that is its more subtle and "peaceful" form. Through the power of spectacle, the powerful relation of images and the human psyche, the screen constitutes us as a globalized mass resonating to scanned pulses, and impulses, of primarily one country. The technological cataract is fully techno-imperialist in scope.

As Marshall McLuhan suggested, that colonization is far deeper than national politics. It is nothing less than a colonization of biology, a reprocessing of the body by technology. In one of his most stark passages McLuhan wrote: "Man becomes as it were the sex organ of the machine world, as the bee of the plant world, enabling it to fecundate and to evolve ever new forms. The machine-world reciprocates man's love by expediting his wishes and desires, namely by providing him with wealth."[27] That process of remaking us into servo-mechanisms of the machine world is a subtle function of electronic technologies, especially television in its guise as the constant provider of spectacle.

The Perfect Machine argues that television has been crucial to the dissemination of that "anti-people climate" in which investment in death-technologies and practices has become the norm. The first part of the book traces the parallel historical development of television and the nuclear industry, both nuclear weapons and nuclear power. The interrelationships reveal the extent to which the mediated image can serve underlying dimensions of which even the patriarchy may be unconscious. As technological cataract, television has been at the forefront of disseminating an ideology of technological omnipotence, the sign of which is surely the bomb itself. By uniting North American society around television, the dominant military-industrial powers subtly united the populace around all technological advance, including the perfection of nuclear weapons.

The book's second part explores some of the particular practices and methods by which the medium is effectively exploited as the new "matrix" of society. The third part, "The Global Agenda", looks at the elimination of borders and boundaries that the medium accomplishes at many levels. This "agenda" is beyond the usual bounds of capitalist imperialism as we've come to think of it. By including patriarchal and technological dimensions, the critique moves into areas that suggest the need for a fundamental rethinking of every aspect of society. As we stand at the brink of Armageddon, that does not seem preposterous to propose.

Part I
The Technological Cataract

2

The New World of the Bomb

In a period when no general ever makes a speech any more without giving God a plug, and self-righteous moralizings ooze from every political pore, real morality has been completely abandoned in our imbecile fascination with these new destructive toys. The atom is our totem; the bomb our Moloch; faith in overwhelming force is being made into our real national religion.

I.F. STONE, The Haunted Fifties

PERHAPS THE KEY word in I.F. Stone's observation is "fascination", derived from the Latin *fascinare*, meaning to charm, to cast a spell. Even though we may be conscious of the planetary annihilation inherent in the new destructive toys, we are spellbound by them, held in thrall by what Stone in 1955 identified as their religious numinosity.[1] Since most of us have never personally witnessed an actual nuclear blast, that fascination with the atom as totem, with the bomb as Moloch, could only have come to us through the media. As radical filmmaker Peter Watkins notes, "The escalation of the nuclear arms race almost year by year parallels the development of television. There's been a very strange synchronism or timing there, actually."[2] It has been television's role to fascinate us with the new world of the bomb, to hold us spellbound as a mass audience in the midst of extraordinary technological advance geared to benefit an elite minority of power-brokers.

□

On December 2, 1942, at the University of Chicago, physicist Enrico Fermi successfully demonstrated the world's first nuclear chain reaction, an experiment conducted in a squash court underneath the university's football stadium. Immediately after this first planned release of nuclear energy, Fermi's superior, Arthur Compton, telephoned the chairman of the Atomic Bomb Committee with the triumphant coded message: "The Italian navigator has just landed in the new world." This landing was obviously a most significant moment in the developing success of the Manhattan Project.

In many ways the U.S. Manhattan Project is the most appropriate model for the military-industrial operations and the complex of technological systems that would come to characterize the "new world" of the postwar era. An absolutely secret endeavour, hidden from Congress and apparently even from the Vice-President of the United States, this massive undertaking dedicated to the creation of the atomic bomb involved thirty-seven installations in nineteen states and Canada, well over 43,000 employees, the first fully automated factories, and a budget of some $2.2 billion at a time of heavy restrictions on spending.[3] A quantum leap beyond any previous business arrangement, this multinational marshalling of expertise from the fields of science, industry, engineering, the military, and public relations – all co-ordinated under the aegis of government expediency – ultimately demonstrated the phenomenal efficiency with which such an organized system could function.

The hiding of this behemoth seems virtually impossible. Yet, from the initiation of the research for the Project in the autumn of 1939 (well before the U.S. entry into the war) until its fateful and spectacular bursts of achievement in the summer of 1945, the Manhattan Project operated in a kind of secret, invisible void. The majority of its thousands of employees had no real idea of the nature of the Project. Only a highly select few knew its actual goal, and even fewer were allowed to grasp the overall complexity of interconnections within the complete Project.

General Leslie R. Groves of the Army Corps of Engineers was in charge of the Manhattan Project and instituted two significant working principles to ensure absolute security: the compartmentalization of information inside the Project, and strict press censorship outside it.[4] While the latter precaution is not surprising, the former needs an explanation, provided by Groves himself seventeen years after the war: "Compartmentalization of knowledge, to me, was the very heart of security. My rule was simple and not capable of misinterpretation – each man should know everything he needed to know to do his job and nothing else."[5]

In the new world of the postwar era, which has its instituting sign in the atomic blasts over Hiroshima and Nagasaki, these two operating principles would remain the framework structuring the smooth advancement of the new technological dynamo. The censorship and manipulation of the press, and the purposeful fragmentation and compartmentalization of information would be crucial to the creation of the nuclear new world.

This framework was already at work in the publicity surrounding the atomic bombings of 1945. As both Vincent Leo and Wilfred Burchett have emphasized, the official press corps sent to cover the results of the bombings focused entirely on two aspects: the visual spectacle of the awesome mushroom cloud in the sky, and the sheer blast-power of the bomb as evidenced by the virtual flattening of both cities.[6] It would take several decades before the U.S. government declassified information and photographs concerning the irradiated bombing victims, who suffered the horrible after-effects of atomic radiation even if they survived the blasts.

This total elimination of the bombings' effect on human beings – a complete exclusion of the vulnerable irradiated human body from public view – shifted North American attention away from lasting corporeal destruction to what Truman, in his press statement of August 7, 1945, called "the greatest achievement of organized science in history".[7] For an unknowing world, the official U.S. government information depicted the atomic bomb in its most abstract (and therefore most technological) form: as a new spectacle of power. Vincent Leo writes:

> Although the Truman announcement did not go a long way in describing atomic weapons, it was the first installment in the "publicity campaign" that was forming around the atomic bombings. As such, it did describe a new basis for American power. It was power guided by the military, in which science and industry participated and which used as raw material the very stuff of the universe. In this arrangement, the nuclear reaction – the physical event – became the mysterious center of power; as if America were able to command the wind and the rain.[8]

In the immediate aftermath, an entire week passed before any official photographs of the bombings were released to the North American press. The intervening campaign of words stressed to North American readers and listeners this new, mysterious centre of power commanded by the U.S. military, science, and industry. When the official photographs did appear, they revealed the bird's-eye view adopted by the airborne military photographer accompanying the bombing crew. As Vincent Leo notes, these official photographs exclude even the city below. They focus entirely on the mushroom cloud expanding out across a timeless sky: a panorama suggestive of an event, a spectacle, devoid of context or even of content.

> The photographs made over Hiroshima and those made in imitation over Nagasaki were almost entirely free of historical complexity; in fact they were almost entirely free of any content at all.... The photographs implied that the weapon was so powerful that it was mistaken for either a cataclysmic natural event or a supernatural one.[9]

These first photographs of the mushroom cloud – taken from an airborne perspective that itself implies transcendence through technology (for, after all, the human body cannot fly) – suggest, as images, the complete irrelevance of human beings in the subsequent unfolding of the new world. Excluding the city (sign of human presence) and celebrating an airborne perspective that defies corporeality, these photographs inscribe the only "eye" which can safely look at this cataclysmic spectacle: the camera eye. Death and living-death below on the ground are not mentioned in these official images. If the photograph of the mushroom cloud has now become the very sign of mass death, it is because we, a public excluded from the plans, have finally caught up with the meaning of events over forty years ago.

The photographic "naturalizing", or even supernaturalizing, of the atomic bombings has its flip-side, its secret equivalent, in the visual material censored by the U.S. government. American forces of occupation sent into Japan after the bombings shot over 85,000 feet of 16mm film documenting the extensive human suffering of the irradiated survivors: footage unavailable for public viewing until 1980.[10] Thus, it was precisely through the eye that full knowledge of this event was fragmented and compartmentalized for the public.

The publicity campaign surrounding this technological achievement – or rather, we might say, the official representation of the atomic bomb to the public – depended on its own splitting of a nucleus: the radical separation of the bomb as technological spectacle (infused with an aura of supernatural power now harnessed by the U.S.) from the bomb as historical cataclysm, the turning point in the designed affliction of suffering on the body at a mass scale. This complete split in the bomb's representation was immediately useful to the manufacturing of public opinion. As Rosalie Bertell writes:

> The U.S. population, although aware that a new and terrible atomic weapon had ended the war, did not receive full details of the human suffering it produced.... Emotionally, most Americans saw the bomb only as a means to bring their sons and husbands home from the war. They never questioned

the illegality of the Manhattan Project, or the lack of Congressional approval of the spending on the bomb or the decision to use it. Few knew enough
of the facts about this atomic attack on human life to protest about the
immorality.... Public opinion was manipulated by the control of information
released by government officials and by the overall elation at "victory".[11]

But this radical separation of the bomb and the human body, this ideological split into two strands of imagery-information to exclude the horrific
and enhance a supernaturalized abstraction of American power, continued well past the immediate needs of postwar "victory" celebration.
Under the sign of its efficiency at ending the war and its spectacular display
of U.S. power in the new world, the Manhattan Project would continue
unabated, though under a new name. As of 1946, its fully marshalled
expertise began to concentrate on expanding the atomic arsenal in a
weapons-building program conducted by the Atomic Energy Commission, which took over the facilities, operations, and technological agenda
of the Manhattan Project. This expansion would similarly depend upon
those two principles adopted and proved entirely effective by General
Leslie R. Groves: the censorship and manipulation of the press, and the
fragmentation and compartmentalization of knowledge and information.
It was within this ideological context that the medium of television was
launched.

☐

Although TV was invented in the late 1920s, its manufacturing and marketing were delayed until the propitious postwar milieu and economic
boom. As early as 1946 the TV receiver set was itself discovered to be a
radiating device, but there is little indication that this was deemed an
important consideration. Certainly, it was not a fact important enough to
scuttle such a potentially lucrative invention. Indeed, it would have seemed
economically unwise to alert a continent of consumers eager for the new
marvels of war-generated enterprise.

In 1937, NBC's Mobile TV Unit was travelling throughout the city of
New York, transmitting live coverage to a tower atop the Empire State
Building for rebroadcast by the main transmitter.[12] Though this mobile
unit was intended to spark interest in the potential of the new medium,
manufacturers had not yet decided to invest in the mass production of sets.
The outbreak of war in Europe in 1939 further impeded TV's marketing
development as U.S. corporations eyed the lucrative possibilities of military contracts.

If the exigencies of war subsequently curtailed the popular dissemination of the medium while hardware manufacturers turned to production
for the war effort, televison as a technology did not go into abeyance

during the war but instead proved useful for scientific purposes. For example, atomic physicist Leo Szilard, at whose urgings the Manhattan Project was instigated, has written of his late-1939 attempts to confirm atomic research theories coming out of France. Szilard describes a significant theoretical breakthrough made while he was working with Enrico Fermi (then at Columbia University) on the still-uncertain possibilities of an atomic chain reaction:

> Everyone was ready. All we had to do was turn a switch, lean back, and watch the screen of a television tube. If flashes of light appeared on the screen it would mean that large-scale liberation of atomic energy was just around the corner. We just turned the switch and saw the flashes. We watched for a little while and then we went home. That night there was very little doubt in my mind that the world was headed for grief.[13]

We have here a hint of the technical interdependence of these two systems. The "screen of a television tube" was necessary to indicate the flashes heralding the "large-scale liberation of atomic energy". At the turn of a switch, the screen conveyed the hoped-for sign to this scientific audience. Oddly, Szilard's description of this moment is suggestive of the television-viewer: turning a switch, leaning back, gazing at the screen, seeing the flashes, and then watching for a little while. Szilard seems to idly observe a spectacle that has no emotional connection to the "grief" he foresees as a result of his work.

A year after that grief hit the Japanese population, RCA (parent company of NBC) got its first mass-produced television sets onto the market, followed quickly by Westinghouse, General Electric, Motorola, and other manufacturers. The date interestingly coincides with another historic event: the beginning of atomic testing by the United States on the Bikini Atoll in the Pacific. Part of this large-scale experimentation was "Operation Crossroads", a series of bomb-tests staged for the world press in 1946. Designed to dispel rumours about the harmful effects of atomic testing, Operation Crossroads (conducted by the U.S. army, navy, and air force) was a massive publicity campaign waged by Joint Task Force One. According to the authors of *Nukespeak* Joint Task Force One "comprised 41,963 men, 37 women, 242 ships, 156 airplanes, 4 *television transmitters,* 750 cameras, 5,000 pressure gauges, 25,000 radiation recorders, 204 goats, 200 pigs, 5,000 rats, and the atomic bombs."[14]

Involving 114 U.S. news correspondents, as well as observers and press representatives from several foreign countries, Operation Crossroads was intended to be "the best-reported as well as the most-reported technical experiment of all time", as the official history of this media-event claims.[15]

Downplaying the hazards of nuclear fallout, these and subsequent series of atomic explosions in the Bikini Atoll were inevitably accompanied by reassuring press releases stating that there were "no visible effects" on living beings exposed to the tests.[16] As in 1945, the publicity focus was centred upon the sheer blast-power of the bomb.

It is interesting that this "technical experiment" staged for the world press in 1946 incorporated television transmitters in its publicity entourage. With set sales barely beginning, there would have been only a few thousand TV owners capable of receiving a television transmission, aside from the fact that, without relay stations, the signal could not have reached the mainland. Thus, we must assume that Operation Crossroads, as a staged media-event, actually served a dual purpose: not only to dispel rumours about harmful radiation, but also to display the new medium of television to the eyes of the world press, and through them, to the world.

This staged "technical experiment", designed to show that there were no visible effects from radiation, replaced what might have been shown to the public: the censored footage of irradiated Japanese. This footage, repressed precisely because of its horribly visible effects, was not a sight in keeping with the goals of the U.S. military-industrial dynamo. By substituting the 1946 imagery of Operation Crossroads for the 1945 footage of Hiroshima and Nagasaki victims, the United States was able to proceed with its nuclear plan. Once again, as with the first photographs of the bomb released to the press, one visual system of signs functioned to exclude another that might have interfered with the manufacture of public consent.

In this sense, television transmitters *had* to be present at Operation Crossroads, regardless of whether or not there was any mainland audience for their transmissions. For a society in which "seeing is believing", the presence upon the Atoll of this latest technological marvel for seeing signified that the most scientific and up-to-date "eye" was being brought to bear on those disturbing rumours about fallout and radiation. When the technical experiment revealed no visible effects on living beings placed at a "safe" distance from point zero of the blast, an unknowing public was reassured.

At virtually the same time as Operation Crossroads was being arranged, Japanese doctors were amassing detailed documentation on radiation-related diseases. This information, like the earlier film footage, was totally censored by the U.S. military. In the late 1940s, when Japanese atomic bomb survivors themselves published a book describing the aftereffects of invisible radiation on their health, it, too, was suppressed by the U.S. occupation forces.[17]

There is a significant contradiction in this. The medium of television (a new technology for seeing or making visible) was launched within a context that necessitated strict secrecy surrounding certain covert operations and knowledge. The U.S. military and industry had embarked on a massive nuclear-weapons expansion program whose experimentation and testing depended on a misinformed public. Yet, in the midst of this secrecy the new medium for informing and making visible was rapidly heralding a new age of instantaneous communication. This apparent contradiction finds its resolution in the specific ideological roles assigned to television in the new world.

Brian Easlea's history of the nuclear arms race, *Fathering The Unthinkable,* raises two significant points about the postwar United States:

> The necessity of a "permanent war economy" which had been first argued as early as 1944 by leading industrialists and military chiefs, was now being vociferously advocated. The survival of capitalism in the United States demanded increased arms expenditure and that in turn necessitated the manufacture of a major enemy.[18]

These twin "necessities" are related to a curious aspect of U.S. media coverage of the bomb in the weeks following the events of August, 1945. *Although no other country had a nuclear bomb,* U.S. radio commentators and writers for prestigious publications like *The New York Times, Life,* and *The Chicago Tribune* depicted vivid scenarios in which U.S. cities were reduced to rubble by atomic bombings. The November 1945 issue of *Life* included a feature called "The 36-Hour War" in which thirteen major U.S. cities were depicted as incinerated by nuclear attack. James Reston wrote in *The New York Times,* "In that terrible flash 10,000 miles away, men here have seen not only the fate of Japan, but have glimpsed the future of America."[19]

In his assessment of this aspect of postwar American culture, Paul Boyer, author of *By The Bomb's Early Light,* writes, "Americans envisioned themselves not as a potential threat to other peoples, but as potential victims."[20] In Jungian terms, this psychological operation is called projecting one's shadow onto the "Other". Rather than face the guilt of what had been done, America projected its own nuclear destructiveness onto some hypothetical enemy: a projection that in turn fuelled the "need" for nuclear weapons expansion. By the late 1940s, television was helping on both the fronts defined by Brian Easlea: naturalizing the "necessity" of a permanent war economy, and manufacturing a major enemy.

A year after TV's presence at Operations Crossroads, where it was both a sign of a new technological "eye" and a diversion from the fallout issues, the medium played witness to another historic event.

In October 1947, the House Committee on Un-American Activities, chaired by Representative J. Parnell Thomas of New Jersey, opened public hearings on "communism" in the film industry. NBC, CBS, and ABC television cameras and microphones were on hand in a caucus room of the House office building in Washington as scores of celebrities assembled under banks of floodlights hung among crystal chandeliers.[21]

In retrospect, it is possible to suggest that the anti-communist hysteria sweeping the continent in the late 1940s was a corollary to the sanitized publicity campaign simultaneously being conducted around the nuclear arms industry. Under the guidance of HUAC Committee members John McDowell of Pennsylvania, Richard M. Nixon of California, Richard B. Vail of Illinois, John S. Wood of Georgia, and Chairman Thomas, the North American public would be encouraged to deflect serious attention from the real (if unperceived) dangers of nuclear radiation, to the perceived (if not entirely real) dangers of communist infiltration throughout the land. If, in the first instance, television could reveal "no visible effects", in the second, it was able to show the scores of celebrities and politicians affected by the "red menace". The deflection was, of course, entirely useful to an expanding nuclear industry.

Broadcasting historian Erik Barnouw has argued that this witch-hunt atmosphere through the late 1940s and early 1950s, coinciding with the formative years for television, made the TV industry "learn caution and cowardice".[22] But the powers within this industry also had much to gain from such a political climate. As Barnouw states, "The National Association of Broadcasters is considered one of the most powerful lobbies, rivalled only by the armaments lobby, with which it has an overlapping constituency."[23] As we know now, some of the major corporate sponsors of TV programming, and some of the first manufacturers of TV sets, were directly involved in the nuclear developments of that time, benefiting from lucrative defence contracts in atomic expansion. General Electric ("Progress is our most important product"), Westinghouse ("You can be sure if it's Westinghouse"), and Du Pont ("Better things for better living") were just a few of the corporations likely to gain from a political climate that was simultaneously hunting down the major enemy in communism and building up "the sunny side of the atom".

Of course, from its inception, there was never any doubt that television's primary function as a mass medium in the United States was to be the dissemination of advertising. Like commercial radio, its programming is the means by which a mass audience is delivered to an advertiser, who pays the network for the systematic delivery of this mass. The creation of this mass audience has always been central to the goals of the medium, but the ideology of this goal becomes apparent in light of the priorities of the late 1940s and early 1950s.

The decision to not demobilize production levels after the war, but to instead continue the permanent war economy advocated by the military and corporate industry, was enhanced by the newly emergent field of motivation research. This field uses in-depth psychoanalytic techniques to probe the consumer psyche. Dr. Ernest Dichter, a pioneer in this new science, announced to the postwar North American business community:

> We are now confronted with the problem of permitting the average American to feel moral ... even when he's spending, even when he's not saving, even when he is taking two vacations a year and buying a second or third car. One of the basic problems of prosperity, then, is to demonstrate that the hedonistic approach to his life is a moral, not an immoral one.[24]

Allowing Dichter's choice of the word "permitting" to resonate in our minds, we can observe that the new medium of television was obviously ideal for solving this "problem". As a continuous visual spectacle in the home, freely provided by the seeming generosity of its advertising sponsors, television could thoroughly demonstrate, hour after hour, this new moral code of consumption – the morality of the hedonistic approach to life – so necessary to the unfolding of the postwar new world.

Thus, if the agenda of the Cold War depended upon a permanent war economy fundamentally based on the expansion of atomic weapons and the nuclear industry, this in turn generated certain socio-political necessities. First, public concern about atomic radiation would have to be dissipated and deflected. Second, a major enemy would have to be delineated as the justification for systematic military-industrial expansion. Third, the public itself would have to be remobilized according to the specific needs of an expanding American dynamo. Clearly, these necessities called for a particular kind of postwar public.

Keeping in mind General Leslie R. Groves's dictate that the security and smooth, unimpeded functioning of a massive system depends on the purposeful fragmentation of knowledge – that "each man should know everything he needed to know to do his job and nothing else" – in the context of postwar nuclear advancement the "job" assigned to the public was specific. Stuart Ewen, in *Captains Of Consciousness,* observes:

> As the Cold War worked its way under the skin of American life, it posed an idealized, consumerized, and increasingly advertised vision of *people as*

basically inadequate. This was at its core. While anti-communist rhetoric never confronted the question directly, consumerized conformity was posed violently against a mode of thought which explored the possibility of people emerging as heroes of their own history. The vision of freedom which was being offered to Americans was one which continually relegated people to consumption, passivity, and spectatorship.[25]

Assigned the "job" of consumption and spectatorship, the North American mass public was to be provided with just what it needed to know to do this, and nothing else. If, implicit in this assigned "job", there is a vision of people as basically inadequate, it is a vision that perfectly corresponds to the new world ushered in by the atomic bomb.

Just as the publicity campaign around the bombings of Japan was built on the radical separation of the bomb and the body, so, too, the new post-war ethos of hedonism could be insatiable and unlimited because it was founded not on the body, which has its natural satiation points, but on the image. It is here, at the site of the excluded body, that the substituted image works. As McLuhan observed, "Media tend to isolate one or another sense from the others. The result is hypnosis."[26] The separation of vision from the other senses, the isolation of the eye apart from the bodily ground, has been entirely necessary for proceedings in the new world. Amputated vision, superseded by TV's technological cataract, allows us to be fascinated by the hedonism of consumption and the pornography of destruction – both of which characterize the postwar era.

3

Atomic Fictions

Technique integrates everything.... The anxiety aroused in man by the turbulence of the machine is soothed by the consoling hum of a unified society.

JACQUES ELLUL, *The Technological Society*

IN 1951 THE UNITED STATES instituted a new phase of atomic-weapons testing. No longer would the bomb-tests occur only on some remote "elsewhere" seemingly far removed from daily North American life. Rather, this new program of bombings was to take place on continental U.S.A., in the remote Nevada desert. Although the Bikini Atoll and nearby Eniwetok in the South Pacific were still used as testing grounds throughout the new phase (receiving a total of sixty-six nuclear blasts during the period 1948-58), for security reasons these areas were deemed unsuitable for the new series of experiments and research planned by the U.S. government and military.

The new program of weapons testing called for the purposeful and systematic exposure of American and Canadian servicemen and personnel to the direct effects of atomic bombs. The army's newly formed Human Resources Research Office needed to gather detailed data on motivation, morale, psychology, and training methods most useful for troops engaging in a nuclear war. Such data could not be gathered by simulations and "war-games", but depended on actually placing troops at the site of nuclear blasts to carefully observe their reactions, both psychological and physical.

From 1951 and on through the decade, between 250,000 and 500,000 servicemen were exposed to the direct effects of atomic explosions in the more than one hundred tests conducted. Countless civilians located in the path of the fallout were also exposed, but apparently the collecting of data on them was of less interest to the needs of the military. Rather, the objects of the study were the men stationed at various distances from the epicentre of the nuclear blasts.

In his book *GI Victims of U.S. Atomic Testing*, Thomas H. Saffer describes his experience of one of the bomb-tests: in this case, a 38-kiloton explosion, test-named "Priscilla".

> Immediately, I felt an intense heat on the back of my neck. A brilliant flash accompanied the heat, and I was shocked when, with my eyes tightly closed, I could see the bones in my forearm as though I were examining a red x-ray. I learned many years later that I had been x-rayed by a force many times greater than a normal medical x-ray. Within seconds, a thunderous rumble like the sound of thousands of stampeding cattle passed directly overhead, pounding the trench line. Accompanying the roar was an intense pressure that pushed me downward. The shock wave was traveling at nearly four hundred miles per hour, pushed toward us by the immense energy of the explosion. The sound and the pressure were both frightening and deafening. The earth began to gyrate violently, and I could not control my body. I was thrown repeatedly from side to side and bounced helplessly off one trench wall and then off the other. Overcome by fear, I opened my eyes. I saw that I was being showered with dust, dirt, rocks, and debris so thick that I could not see four feet in front of me. I could not locate the person who had been nearest to me in the trench. A light many times brighter than the sun penetrated the thick dust, and I imagined that some evil force was attempting to swallow my body and soul.[1]

In some of the 1950s bomb-tests in Nevada, troops were stationed as close as two miles from ground zero, then ordered to proceed directly into the epicentre after the explosion. These troops were not issued any special protective clothing, and only in some cases were they provided with gas masks or goggles through which to watch the fire-ball mushrooming into the sky. Given the Japanese medical documentation censored by the military in the late 1940s, it is likely that the effects of radiation were known to the military experimenters of the 1950s if not to their human guinea pigs. These possible effects may have been ignored because they were not immediately significant to the purpose of the research: which was to determine the most effective combat procedures following a nuclear explosion.

Such suspicions are enhanced by two facts. First, the military scientists themselves were fully suited-up in protective gear before entering the zones where troops had been stationed. Second, those human guinea pigs of the 1950s have subsequently found it extremely difficult to locate military records proving their participation in the experiments. Both factors call into question any seeming ignorance or innocence surrounding the experimentation.

This whole new phase of nuclear testing obviously called for a renewed public-relations campaign that would have to deal with two new dimensions: (1) bombing the homeland, and (2) exposing servicemen to the blasts. Unlike Operation Crossroads, staged for the press in 1946 at far-off Bikini Atoll, this new campaign would have to be pitched directly to the American public in whose backyard the bombs were exploding.

Given these factors, it should not surprise us that television was a primary means through which this public-relations effort was conducted. A popular mass medium already installed in millions of North American homes by 1951 and expanding exponentially across the continent every month, television was the obvious site upon which to build a society unified around the bomb. The very popularity of this mass medium, with its pleasurable entertainment function in the home, could "rub off", so to speak, onto the spectacle of bomb-blasts imaged on its screen.

As the tests proceeded throughout 1951 and into the spring of 1952, the Atomic Energy Commission decided to invite selected members of the American press and television to observe and play witness to this new development in nuclear experimentation. On April 22, 1952, during the Tumbler-Snapper series of bomb-tests, the Commission permitted the first television coverage of the blasts. This 1952 premiere marks the conjoining of the two dominant mass media of our time: television and the bomb.

There were, of course, ripples of protest from the public about the use of troops in the experiments, but the Atomic Energy Commission handled the situation with aplomb, its public-relations machine hastening to grind out reassurances about the safety of the manoeuvres. The U.S. Advertising Council also stepped in to assist in 1952, sponsoring a telecast of one of the bomb-tests at Yucca Flat. The Advertising Council used the occasion to emphasize for the mass viewing-audience the so-called "safe threshold" of radioactive fallout, the wisdom of erecting home bomb-shelters, the need for a Ground Observer Corps, and the ever-present communist menace.[2]

At first glance, it would be easy to pass over this telecast as entirely ordinary and mundane: the U.S. Advertising Council sponsors a telecast of the atomic bomb exploding over Yucca Flat, Nevada. It seems remarkable, even somewhat banal: a sponsored telecast, just another program, within an evening of entertainment. But it's this very banality that we must focus on. One minute you're watching, say, *I Love Lucy*; then it's *The Life of Riley* followed by the *Yucca Flat Bomb-Test* followed by *Your Show of Shows*. The nuclear telecast fits nicely into the flow of programming.

This is not to detract from the intelligence of the viewing-audience. It is, rather, to underscore two significant characteristics of the television medium. First, at the level of the transmitted signal, each image is equally fascinating. The medium has the effect of flattening (or inflating, if you prefer) the significance of everything on it onto a single plane. Second, the medium is one of containment. This is television as the new "matrix", the new "ritual container" for a society otherwise devoid of meaningful face-to-face rituals and social institutions. Although in the early 1950s television was still amassing its institutional power, as a technological screen it was already framing an ideological vision. The television apparatus, in comparison to its imaged content, had become the "higher logical type", to use the terminology of Gregory Bateson.[3] Quite literally, TV provides a neat framing box around what it shows us, but in a larger sense it contains (in the sense of managing, subduing, encompassing) the sights it provides. Both these aspects of television containment – as new "matrix" or "ritual container", and as the techno-institutional frame for programming – are entirely significant to the telecasts of the bomb-tests in the early 1950s. By televising them as a sponsored program, they became banal, non-threatening, not significantly more remarkable than other telecasts.

In a 1950s milieu of almost total mystification about the bomb, it would be easy to overlook the amazing configuration of elements converging in such telecasts: the greatest weapon of mass death becomes a sponsored spectacle on the greatest medium of mass entertainment. Even leaving aside the radiating properties of the TV set at that time, we sense the terrible ironies at work. Spectacle and spectre converge on the screen. Mass amusement and mass death meet at the site of our new form of sight. The technological cataract can fuse them both. The "job" assigned to the mass postwar public – consumption and spectatorship – takes on decidedly nuclear overtones.

By 1952 the nuclear industry had also advanced to a new stage of nuclear weaponry: the hydrogen bomb – a fusion device far more powerful than the atomic bombs dropped on Japan. This terrible step in mass destruction called for further mystification of the mass public, in line with the security dictates previously advocated by General Leslie Groves. As Judith Randal writes:

In May 1953, when the public was complaining increasingly of the dangers of nuclear testing, the Atomic Energy Commission issued a memo saying that the Commission press releases and speeches should not use the words "thermonuclear", "fusion" and "hydrogen". The memo asserted the President [Eisenhower] says "Keep them [the public] confused as to fission and fusion."[4]

As part of this purposeful confusion, the nuclear establishment shortly thereafter initiated its publicity campaign around "the peaceful atom". Just months after his apparent directive to keep the public in the dark about the hydrogen bomb, Eisenhower delivered his "Atoms For Peace" speech to the UN General Assembly. In the subsequent publicity push for the development of nuclear reactors, television again played a central role.

During the early 1950s, U.S. Signal Corps cameramen shot more than 100,000 feet of footage of atomic energy installations – footage that was made available to TV producers and broadcasters on the agreement that they would submit scripts to the U.S. Atomic Energy Commission explaining in detail how they planned to use the footage. By controlling access to the installations, as well as by establishing this script-review procedure, the AEC fully determined and sanitized the TV portrait of atomic energy throughout the 1950s and afterward.[5] In addition, GE, Westinghouse, and the AEC produced and provided hundreds of free, pro-nuclear films to domestic television stations throughout the 1950s and 1960s. Even the U.S. Chamber of Commerce joined the campaign, sponsoring the television film *The Atom Comes To Town* in 1957.[6]

But television in the 1950s did more than dutifully broadcast P.R. films for the industry. It also staged its own nuclear pseudo-events. For example, on September 6, 1954, President Eisenhower was called upon to ritualistically inaugurate construction of the first commercial nuclear reactor in the United States, to be built in Shippingport, Pennsylvania. The form of this inauguration as a P.R. event deserves our scrutiny. Standing in a Denver TV studio, Eisenhower was asked to brandish a radioactive wand over a counting device. "When the counter's needle swung across the dial, it electrically set in motion, 1,300 miles away at the Shippingport, Pa., plant site, an automatically controlled power shovel which scooped up the first symbolic shovelful of earth."[7]

In this televised pseudo-event, we find another significant configuration of elements. First, the President becomes the trusted signifier of American democracy, literally conferring upon the proceedings not just the authority of his office, but all it stands for in the political process. But he is also more. He holds a wand, like a kind of magician about to perform a trick. It, too, is more than a wand. Rather, it is a fully magical wand of radioactivity that has a noticeable effect on the counting device over which it is brandished. The interaction of these two devices – or should we say three: wand, counting device, and President – sets in motion another device, separated by 1,300 miles but somehow magically linked to them. An automatically controlled power shovel, triggered by this magical act, scoops up the first shovelful of earth for what is destined to become the first U.S. commercial nuclear reactor, the last device in this series of interacting devices. In retrospect, all this technological magic in the staged P.R. event was seemingly necessary to deflect whatever rational worries might be lingering in the public about radiation.

There is one device in this complex of devices that is less easy to see. Like a magical trick of its own, the television apparatus makes visible this sequence of technological magic, but is itself transparent and disappears from view, because we are looking through it. Although the entire magical act is being initiated inside a Denver TV studio where Eisenhower waves his wand, and although the conveyance to us of the wand's magic depends on TV's ability to cut to a location 1,300 miles away to show the result, we are not intended to consider television as part of this sequence of interacting magical devices. Instead, we are meant to look through TV as though it were simply a replacement for our eyes. TV is thus the better magician, something beyond all the other magical devices it shows, because it stages its own disappearing act while it reveals the workings of other technologies' magic. Indeed, the entire series of events unfolding in this public relations moment seems to imitate the larger magic of TV.

But this time it is the TV-viewer who, in a "presidential" role, flicks the switch and inaugurates the electronic magic which brings, over vast geographical distance, an image that automatically appears on the screen. Like the President, who is necessary to the ritual aspects of this 1954 pseudo-event because of the aura of his office, the TV-viewer is also necessary to the television ritual. The office of our assigned job as consumers and spectators also confers upon the proceedings a certain authority. But, like the President in this series of interactions, the TV-viewer is one device among several interacting in a sequence of technological magic.

By staging this nuclear P.R. event for television, the growing nuclear establishment of the 1950s blessed the initiation of the first commercial nuclear reactor with all the popularity and trustedness of the television medium – which had, by 1954, penetrated into more than half the households in North America.

Yet, as with the mushroom cloud photographs of the 1945 bombings, the official televised imagery of the inauguration of nuclear generators would also exclude a certain strain of knowledge and information. For instance, the mass television audience was not informed of the first major accident at a nuclear reactor, which took place well in advance of this 1954 staged TV pseudo-event. On December 13, 1952, a hydrogen explosion occurred at the NRX reactor in Chalk River, Ontario. The explosion killed one man and seriously contaminated five others, while largely destroying the reactor core and releasing one million gallons of highly radioactive water. Among the U.S. Navy technicians sent to assist in the emergency was the young Jimmy Carter. Dr. Rosalie Bertell writes, "The accident was shrouded in secrecy in the United States, Canada and Great Britain for the sake of national security. Understanding the implications of nuclear accidents might reduce civilian co-operation with the growing weapon industry."[8]

For this reason as well, the mass public was not informed during the hype surrounding the "peaceful atom" that nuclear reactors generate more than electricity: that they are actually necessary to making weapons. After several months, the "spent" fuel rods are reprocessed to yield the fissionable uranium and plutonium needed for nuclear weapons. In place of such information, television in the 1950s substituted an image of the nuclear establishment inbued with technological efficiency and magic: heightened, no doubt, by the wonders of seeing it all on the television medium itself.

□

Just at the point where television had become a fully *mass* medium through the scope of its reach by the mid-1950s, a curious change occurred in broadcasting practice. Using the language of J. Robert Oppenheimer, it appears that what was originally "technically sweet" about the new technology of television was its totally unique ability to transmit images and sync-sound live, in real-time, across great geographical distances. Bypassing such time-consuming stages as film processing and film editing, TV's instantaneous electronic transmission of live coverage heralded a breakthrough in space-time parameters unlike anything previously achieved by technological means.

Yet, less than ten years after the start of commercial TV broadcasting in the United States, this unique capacity of the medium had been largely eliminated from conventional broadcasting practice. By the mid-1950s, 80 per cent of network broadcasting schedules consisted of programming produced on film.[9]

If the medium of television had initially been launched primarily as a new means for distributing the older medium of film, we might have no reason for curiosity about this turn of events in television practice. But, given a highly successful initial period of live transmission, during which TV's popularity spread from a few thousand viewers in 1946 to more than half the nation's households by 1954, this change from live transmission to filmed broadcast seems not so much some "natural" evolution, but a particular historical decision made by industry powers for specific ideological reasons. Apparently, by the early 1950s, what was technically sweet about the new medium of television had become politically distasteful.

Of course, we might speculate that live transmission, especially in the early days before the full packaging of live pseudo-events, is the technological equivalent of the Freudian slip. At every turn, the failure of the facade can occur. Someone could inadvertently state the wrong thing and suddenly the carefully orchestrated facade tumbles before our eyes. In the paranoia of the early 1950s Cold War, such considerations are not outlandish. Indeed, it has been suggested that the live broadcasting of the Army-McCarthy hearings, while very popular with viewers, was a significant factor in scaring the TV industry away from live transmission. An audience of some thirty million TV-viewers was beginning to sense that the whole Reds-Under-The-Bed theme they'd been hearing since the late 1940s was somewhat suspicious.

But the change in the mid-1950s from live transmission to filmed product for broadcast had its greatest, and perhaps most interesting, repercussions in the area of U.S. TV drama, the popular fictions of prime time. In this genre especially there was apparent a subtle ideology behind the switch.

In the early years of network TV a single sponsor usually hosted an entire program, a practice that allowed the popularity of a given show to accrue to the advertiser. At the same time, there were two major strands of popular programming in prime-time drama. The first was the live episodic series: a narrative form borrowed from commercial radio, with continuing casts of characters and recurring sets, and a self-contained dramatic story resolved each week. Live TV programs like *Man Against Crime* and *The Aldrich Family* followed this episodic form.

The second strand of live prime-time drama was the anthology series in which an original teleplay, with unique cast, plot, sets, was broadcast live each week. By the early 1950s there was a long list of live anthology drama series, including *Philco Television Playhouse, Goodyear Television Playhouse, Kraft Television Theater, Studio One, Robert Montgomery Presents, U.S. Steel Hour, Omnibus, Motorola Playhouse,* and *Playhouse 90.* These programs were attracting not only millions of viewers, but also a host of creative talent. Writers such as Reginald Rose, Rod Serling, Paddy Chayefsky, Robert Alan Aurthur, Horton Foote, and Gore Vidal, and directors like Delbert Mann, David Susskind, Sidney Lumet, John Frankenheimer, and Arthur Penn seemed to find, in the live anthology series, a narrative form and a unique TV aesthetic that appealed to their talents.

Unlike film, live TV drama was primarily studio-bound, with its bulky electronic cameras and its switching panels for the instant editing that characterizes live transmission. Thus, live TV drama meant relatively small studio sets and a reliance upon close-ups and medium-shots to tell a story. The physical confines ruled out panoramic action on a large scale, and the fact of live broadcast eliminated tricky action stunts that might not work the first (and only) time.

But this live TV aesthetic enhanced the possibilities of constructing dramas around other concerns, especially the seeming constraints of the close-up. As broadcasting historian Erik Barnouw notes, "The human face became the stage on which drama was played."[10] The resulting teleplays highlighted psychological realism, inner emotional conflict, and what Paddy Chayefsky called "the marvelous world of the ordinary". In live anthology drama, as Barnouw says, "The action was not physical action, it was psychological confrontation. These were tight, indoor dramas: psychological dramas with sociological overtones."[11]

These very elements contributed to conflict with the powers of the TV industry. Advertisers wanted programming that would enshrine consumerism and envy as a way of life. The dramas of the ordinary tended to display none of that hedonistic dimension necessary to the new moral code advocated by Dr. Ernest Dichter and his colleagues. Indeed, the psychological depth in live anthology dramas tended to make the commercials seem fraudulent.[12] The ads were proposing, after all, that any problem could be fixed by the purchase of a product. In addition, these dramas with their sociological overtones tended to locate the sources of dramatic conflict in the depths of American life itself: implying the need for socio-political change, awareness, and the raising of controversial issues. As Rod Serling stated at the time: "Of all the media, TV lends itself most beautifully to presenting a controversy. You can just take part of a controversy and, using just a small number of people, get your point across."[13]

Thus, live anthology drama of the early 1950s contained the seeds of a threat to the status quo, not just in raising controversial issues, but in the narrative form itself. The programs presented nonglamorous characters who achieved a certain psychological depth by going through substantive change in the face of socio-political problems. This narrative form assumed a television viewer engaged with the problems and controversies of the times and looking to dramatic constructions to further illuminate the process. This, after all, is what traditional drama has always done.

During the 1951-52 season, two new episodic series appeared on the scene and offered advertisers an appealing alternative to sponsoring live anthology drama, with all its controversies and discontents. *I Love Lucy* and *Dragnet* quickly demonstrated the popularity of the episodic narrative form, but this time produced on film. The filmed episodic series held out a number of attractions to sponsors. By being able to shoot retakes of scenes, producers eliminated the potential errors of live transmission. By using the same characters and sets week after week, the form ensured not only that production costs would be lower overall, but also that once a safe programming premise had been created, the episodes could be produced almost automatically, with little danger of controversy sneaking in. As well, the filmed episodic series could be designed to nicely match the commercials and provide a more suitable context for them: one that did not make the ads seem fraudulent in any way.

The filmed episodic series also posited a different kind of viewer than that assumed by live anthology drama. The filmed episodic series focused on *unchanging* characters whose behaviour remained predictably familiar week after week. Once the characters were established, this narrative form eliminated the dramatic notion of character development. The weekly reiteration of fixed characterizations became part of the viewing pleasure. As a narrative form, the filmed episodic series excluded not just psychological depth, but also a dramatic structure in which change was possible. Once the episodic premise had been established, nothing about it could change except for the weekly variations on *external* threats to its status quo.

In the situation comedy, these threats were implicit in the superficial misunderstandings or mixups that had to be resolved in the course of the half-hour, restoring the original premise of the series so that the program could return the following week. In the crime series, the weekly threat to the status quo of the series was embodied in the criminal deviant, the externalized "other", whose capture or killing neatly resolved the narrative and returned it to its pristine state so the next week it too could start again from the same premise. In both types of episodic series (sitcom and crime), the narrative itself demonstrated the ease with which the status quo remained sacrosanct: beyond any real change.

Implicit was the view that self and society were completely unchanging, with change itself located only on an external plane that matched the product variations introduced by sponsors and marketing. Problems never derived from the real conditions or contradictions of society, which would have called for substantive socio-political change. Instead, problems sprang from a superificial or externalized source of trouble that could be neatly resolved in the course of the drama. As Barnouw observes:

> On these programs drama never derives from problems within a central character or characters. The viewer is never asked to look for problems within characters with whom he mainly identifies. The trouble is always someone else – never oneself. Introspection is never encouraged. The pattern has ritual appeal for the immature; ceaselessly re-enforced, it may prolong immaturity. It also holds seeds of paranoia.[14]

Such a world-view provided a more suitable context for TV ads than live anthology drama. Sponsors quickly abandoned the live drama, making a wholesale switch to filmed episodic series by the mid-1950s. As David Susskind, former producer-director for *Studio One* and *Kraft TV Theatre*, stated:

> When the discovery dawned upon the key people that television was the greatest merchandising and advertising instrument ever invented in this country, it was then taken hold of bodily by the businessman.... Innovators were driven out and hard-hearted, hard-headed statisticians, salesmen and business took command of the industry.[15]

But the fact is, the TV industry in the United States had been launched in 1946 by just such businessmen and precisely for this advertising purpose. Thus, the wholesale narrative switch must be explored in connection to the larger socio-political milieu of the early 1950s.

Given the U.S. media paranoia that immediately followed the bombings of Japan – that weird turnabout in which the United States was perceived as the potential victim of the atomic bomb – a TV narrative structure that highlighted an external threat might serve to (oddly, insanely) justify the fact that the United States had become the victim of its own bomb: with Nevada bomb-tests occurring as often as once a week towards the latter part of the decade. If such bombings were seen to be entirely necessary because of an external, foreign enemy, then the nuclear build-up could proceed. Similarly, in its elimination of the notion of character development, psychological depth, any change, the filmed episodic series posited and celebrated the end of the individual person meaningfully involved in a changing world. It was now a fixed world, beyond real change other than in its external products and external variations on a theme: a fixed world peopled by fixed characters who weekly reiterated the rigidity of an overall framing premise that remained sacrosanct. Like cogs in a machine, the characters in such series returned again and again to situations that differed only in the sense of posing a new plot but with the underlying structure and premise unaltered from week to week.

This narrative switch signalled an extreme narrowing of narrative possibilities. As the conventions of the sitcom and crime series became entrenched, TV drama congealed and hardened around fewer ideas, structures, and creative avenues. As Susskind said, innovators were driven out. But innovation itself was excluded from the very notion of episodic series, whose rigid framing premise ruled out the introduction of anything that could significantly alter it. At the simplest level, characters in an episodic series from the 1950s into the 1970s did not learn or change as a result of their experiences or insights. Rather, they seemingly underwent amnesia week by week, falling back into the habitual patterns and traits assigned them by the ruling premise of the series.

□

This wholesale switch in TV narrative practice meant that henceforth the medium would reiterate a narrowing of the possible. Although the previous era of live broadcast had its own ideological constraints, there was nevertheless an increasing rigidity in both the change from live broadcasting to filmed product, and from anthology drama to formula series. This rigidity and narrowing of the possible into a fixed, unalterable "world" coincided with a Cold War milieu in which, as Stuart Ewen said, the vision of freedom being offered to people had diminished to consumption, passivity, and spectatorship.

At the same time, in the real world beyond the television screen, the framing premise of American society was itself congealing around the needs and goals of the nuclear industry – an industry predicated on the supremacy of "unalterable" systems and structures comprising the expanding war-machine. The U.S. and Canadian servicemen stationed in the path of the bomb-tests had no individuality (and, one might add, no nationality) other than as statistical data in the experiments and research. Their individual characteristics and personal histories were insignificant to the mass role assigned them: statistics in a mass profile useful to nuclear combat procedure. In the postwar new world, the needs and interests of the individual are collapsed into the greater needs of the prevailing technological systems. Those systems frame the possible, even establish, in Noam Chomsky's phrase, "the bounds of thinkable thought".[16] As a technological system, television had found, by the mid-1950s, the formats for spectacle most appropriate to its role.

But let's go a bit further. We now have some thirty-five years' worth of examples of American sitcom and crime series formulas from which to derive insights into an underlying ideology. These two formulaic premises have been the mainstay of prime time ever since the industry made the narrative switch during the Cold War. Strangely, or perhaps not so strangely, what emerges from a scrutiny of these formulas is that they are closely interconnected: the sitcom and crime show build on each other, deriving added meaning through combination (which, of course, is how they are broadcast in a schedule's flow). In many ways, the gaps in one formula are filled in by the conventions of the other. Together these two dominant formulas reiterate a world-view or zeitgeist that has some very telling aspects.

We can begin simply by looking at their titles. Crime series usually have titles that use the last name of the main character, a way of conveying authority, formality, and power: *Quincy, Kojak, MacGyver, Mannix, Cagney and Lacey*. By contrast, sitcoms usually have titles that use the central character's first name, suggesting informality, intimacy, relaxed familiarity: *Lucy, Rhoda, Alice, Kate and Allie*.

Another grouping of crime series uses titles that refer to the exterior urban milieu, a milieu central to the formula: *Miami Vice, Hill Street Blues, Vegas, NYPD*. Sitcoms, however, rarely have titles that refer to urban setting. Instead, the titles often indicate the importance of the family unit: *Family Ties, Three's Company, All in the Family*. There are occasional exceptions to these conventions of titling, but usually the name of the program signals which formula is at work in a specific series.

Neither formula is broadcast live, but crime series tend to be shot on film, while sitcoms are shot on videotape. The choice contributes to the "look" of each formula. The filmed image is considered to be more crisp and to have more depth-of-field than the videotaped image, so objects and movement in both foreground and background of a shot are more clearly visible. The sharpness of image also contributes to the "hard-edged" qualities that one expects from the formula: with stark outlines of people and objects, definite areas of shadow or black, contrasts in size between buildings and people, contrasts between foreground and background. An image-in-depth is also important for the kind of action that takes place in the crime formula. Movement towards the camera is more dramatic than lateral movement across the screen, so action in this formula tends to travel from background into foreground to create a sense of urgency, especially when a chase is involved. There is a larger scope of action in the crime series, and the greater depth of the filmed image adds to it.

Sitcoms, however, are usually videotaped, yielding an essentially flat image without much depth-of-field and softer in its outlines. This type of image suits the predominant shooting style and movement of characters in sitcoms. Camera distance is typically medium-shots and close-ups, and camera movement is greatly restricted. This shooting style contrasts with that of the crime series, where extremes of camera distance are common (from extreme long-shots to extreme close-ups) and camera movement is more elaborate. Because sitcoms are videotaped on the proscenium stage of a studio set, neither camera movement nor actor movement can be large-scale. Sitcoms are generally characterized by restricted movement, smaller gestures, and extensive use of facial reactions.

The sitcom formula almost never uses location-shooting, apart from the introductory "signature sequence" which establishes an exterior milieu that is then never again seen in the course of the given episode. In the crime series, though, urban milieu is central, and location-shooting exploits the features of the city as much more than a backdrop. The city is central to the recurring theme of the crime series formula: the power struggle between forces of law and lawlessness. Whether the main character is male, female, mutant, or technological (as in *Knight Rider* and the bionic programs), he / she / it is the embodiment of the power of law, the principle that backs up action and motivation. Using quick, effective gestures and taking decisive action, the characters in this formula are forceful figures of authority. Even Raymond Burr's *Ironside*, confined to a wheelchair, is suitably hard-edged. Of course, within the power struggle being played out in urban settings, the forces of lawlessness are similarly decisive. This formula, then, is populated by a set of conflicting figures of power on one hand, and a set of vulnerable bystanders and victims on the other.

These conventions are in obvious contrast to the sitcom, where the central recurring theme is the necessity for compromise and understanding in relationships. The daily, mundane frustrations and human problems associated with family units or close-knit social groups provide the basis for story-lines and jokes in this formula. This preoccupation is mirrored by the interior, familial sets used. It is personal, not public, space; just as the conflicts are seemingly personal, rather than social, in scope. When the situation is the workplace, as in so many 1970s and 1980s sitcoms (*Mary Tyler Moore, Barney Miller, WKRP In Cincinnati, Cheers, M*A*S*H, Night Court*), the seemingly public space is domesticated and made familial by the close, personal interactions among the characters, who function like a family unit, and by the personalized touches provided to the recurring interiors.

Sitcom characters, however, aren't particularly decisive or effective. In fact, much of the humour is based on their ineffectiveness as they struggle to solve a problematic situation, or as they vacillate among various ways of dealing with a conflict. The characters in sitcoms are meant to be endearing precisely because they are unsure of themselves, ineffectual, indecisive, or mistakenly forceful. Indeed, when characters are forceful they often learn some lesson in compromising in the course of the half-hour. The lesson is, of course, forgotten by the following episode.

In the mid-1970s TV critic John Leonard sat in on discussions with TV writers and found that they operated on a basic assumption: "The public doesn't like to feel that people on TV are better than they are." According to Leonard, an important trait of sitcom characters is vulnerability: "To a certain extent they have to lose week by week. There is a kind of 'Hollywood-Universal City-TV sitcom-abstract principle' that to be funny, you can't be wise. I mean wise in the sense of wisdom, not in the sense of being a 'wise-acre'. You can't be wise, you have to have somehow stumbled through to the right solution so that you are not presented as being better than anyone else."[17]

This sitcom principle contrasts with the crime series formula, where the central character *is* wise, ahead of the story, forcing the action. Even *Columbo*, or Louie Ciccone in CBC's *Seeing Things*, fits this convention. In *Seeing Things*, Louie's "visions" keep him ahead of the story, quite literally.

Unlike sitcom characters, the main characters in crime series (with certain notable exceptions) are rarely what one could call "family types". Marriage, children, the family unit are, for them, either something in the past or an implied hindrance to their "lifestyle": which mainly revolves around work. In fact, in this formula, work is the *raison d'être* – the central personal commitment. Time-off may be alluded to, but it is rarely shown and doesn't take up much screen-time. When it is an integral part of an episode, it becomes the occasion for crime-solving, as though the crime series character simply cannot escape the necessity for work. Thus, the hour-long crime show focuses almost entirely on work: the solving of a crime, usually done effectively and without compromise. It is a formula peopled by workaholics functioning in the public sphere who have little, if any, private life themselves. This is the classic description of the formula ever since *Dragnet*. The exceptions such as *Cagney and Lacey* tend to prove the rule.

By contrast, in the sitcom formula, when the job does provide the situation for the comedy (as in most recent examples), work is used to reveal frustration, daily problems, small compromises. The characters aren't the dedicated and successful workaholics found in the crime series. Furthermore, the job provides a substitute family unit for the sitcom characters, while the crime series' central character usually works alone or with a single partner, but at a job that fully separates them from the pack.

In its sound effects the sitcom depends on a "laff-track", which structurally corresponds to the music-track of the crime series. Music is never used as background in a sitcom, but it is an important element in the crime series. Interwoven with the sounds of squealing tires, shouts, or gunshots, the music-track helps establish mood and pace and is used to underline dramatic moments. Not only are there often layers of foreground and background sound, but usually the level of the music-track is also brought up into the foreground to heighten a dramatic climax: a kind of audial depth in keeping with the depth of the filmed image.

Another audial difference is found in the kind and amount of dialogue. Sitcoms generally use rapidly paced and plentiful dialogue, with a verbal joke having at least one topper. Canadian director Norman Campbell, who has worked on sitcoms in the United States, argues that dialogue determines the "cutting" of a program: "I could almost look at a script once and tell what the camera cutting would be on it because if you're dealing with comedy, you can see where the jokes are, whose face you want to be on, and when you want to cut to a reaction."[18] In the sitcom, cutting enhances dialogue.

Crime series, because of their action orientation, usually involve far less dialogue than a sitcom and cutting is used to highlight movement. In their most classic examples, the energy of the sitcom is located in the soundtrack through dialogue and laughs, while the energy of the crime series is located in the visual-track through movement and camera-work.

Structurally, many of these formulaic conventions appear as a set of oppositions:

Crime Series Formula	*Sitcom Formula*
filmed	videotaped
sharp, in-depth image	soft, flat image
hour-long	half-hour long
large-scale movement	restricted movement
exteriors, location shooting	interiors, studio-bound
cutting enhances movement	cutting enhances dialogue
music-track	laff-track
public space	personal space
urban setting, city streets	familial setting, rooms
loner "hero"	family or group "heroes"
successful work	frustrating work
uncompromising solutions	compromising solutions
powerful figures	powerless figures
formal, heroic	informal, nonheroic

Using such a structural grid, we can sense the ways in which some TV series (for example, *Moonlighting* and *Miami Vice*) play with, or go against the grain of, the classic elements assigned them. After some thirty-five years of familiarity with these two formulas, in the 1980s the industry went into a period of stylistically self-conscious production that assumed a certain degree of cross-breeding between formulas.

But if the above elements characterize the two formulas that have long predominated on prime time, it is also true that the formulas interconnect and fill in each other's gaps. There is a curious collusion between them, which raises the question of an underlying ideology for the deep structure of American prime time.

Years ago, vice-president of ABC programming Bob Shanks expressed the production premise at the core of U.S. television:

> Program-makers are supposed to devise and produce shows that will attract mass audiences without unduly offending those audiences *or too deeply moving them emotionally*. Such ruffling, it is thought, will interfere with their ability to receive, recall, and respond to the commercial message. This programming reality is the unwritten, unspoken "gemeinschaft" of all professional members of the television fraternity.[19]

In the two central prime-time formulas, wisdom, authority, decisive action, and leadership qualities are reserved for figures set apart from the crowd and enacting a power struggle around the law: law-enforcers and law-breakers nightly re-enact their decisiveness and expertise on the screen. Their milieu is the outside world, the public sphere beyond the confines of the living-room. It is seemingly a sphere of danger, crime, disaster, and complexity that only a select few can cope with or even understand. Uncompromising and hard-edged, these few distinguish themselves by their authoritarian wielding of power in service to, or against, the law. Formal, heroic and singular, the law-enforcers roam and rule the public sphere whose scope and large-scale dimensions are not intimidating to them.

By contrast, the rest of prime-time's figures, who tend to more closely resemble the mass-viewer watching at home, must fumble through their small, daily frustrations, content with a restricted and private sphere of action. Rarely seen to enter the exterior, larger world of society, these nonheroic figures are confined to familiar space and devote themselves largely to seemingly privatized personal problems that appear to have no reference to, or source in, larger social conditions or context.

There emerges, then, a definite split between the public sphere and the private realm: a split that is further characterized by extremes, with danger and crime associated with the outside world, and safety and humour associated with private space. This split would suggest that the public sphere and all it entails – whether governmental rulings, corporate decisions, environmental concerns, or political action itself – is virtually unconnected to "private" frustrations and well beyond the understanding or influence of ordinary people, who apparently lack the authority and expertise to deal with such complexity.

These basic, underlying premises are assumed to be not unduly offensive or too deeply moving emotionally for a mass audience. Indeed, these premises are perhaps central to constructing the mass in the first place. By denying action, influence, and leadership in the public sphere to all but a select few, the prime-time zeigeist reiterates the diminished role assigned to the mass population in the new world. Any attempt on our part to demonstrate competence, authority, and effectiveness beyond the privatized realm of purchasing-power and spectatorship would suggest not only that we have assumed we are better than others, but also that we have stepped out of our "natural" place into the "ruthless" public arena.

When we connect this underlying ideological premise with other aspects of the filmed formula series — the elimination of character development, psychological depth, and real change — we see the extent to which the wholesale narrative switch in mid-1950s Cold War television practice mirrored, and reinforced, the larger prevailing ethos of technological supremacy and authoritarian rule by an elite cadre of experts and enforcers.

Here we must return to that historic telecast of the Yucca Flat bomb-test, arranged by the U.S. Atomic Energy Commission and sponsored by the U.S. Advertising Council. This time let's focus on the agencies involved: one, the power behind vast nuclear expansion, the other the official representative of the entire advertising industry and thus the power behind television itself and much of American existence. The joint effort of these two agencies in the telecast signalled that, at a deep level, their goals and ideologies were similar. In this historic moment, the greatest medium of mass spectacle and the greatest medium of mass death were conjoined in the TV image of the mushroom cloud: technological cataclysm seemingly contained by technological cataract. The twinned and intertwined necessities of both media in a society increasingly based on unfreedom make it difficult to say which provides (in the words of Ellul) the "consoling hum" for the other.

4

The Sweetening Machine

THE NARRATIVE SWITCH in U.S. TV practice in the mid-1950s is only one aspect of this turning point in the industry. Other aspects are equally important, especially in terms of altering both the business structure and the mode of production for the medium. It might be useful to here retrace that mid-1950s turning point from a slightly different angle. If what had originally been "technically sweet" about the medium – its ability to transmit images and sync-sound live, in real-time, across great geographical distance – had become politically distasteful by the end of the medium's first decade, then that distaste encompassed more than narrative practice.

In the era of primarily live broadcast, a given show had a half-hour or hour-long existence, after which it simply disappeared into the ether, so to speak. The bulk of programming was thus beyond playback, unrepeatable. Although some shows were stored (for archival purposes) on kinescopes – made by pointing a movie-camera at a TV screen during the live broadcast – most transmissions simply lasted for the duration of the show and then were gone. Undoubtedly, this was part of the excitement and appeal of live broadcasting. This fact coincided with another one. In those early days, the U.S. networks produced most of the shows themselves, in-house (as they say), and sold sponsorship of an entire show to a single advertiser. It was a kind of enclosed, self-contained operation: the networks produced their own shows with their own paid staff, financed production by selling ad-time, the shows were broadcast and disappeared, then the networks produced another batch of shows the next day. Simple.

But then the U.S. movie industry – which at first had been very threatened by television because the new medium was keeping people at home and out of the cinemas – began to realize that it just might be able to get a nice piece of the action. If, movie people reasoned, instead of being produced live, TV programs could be produced on film and *then* broadcast, all those shows wouldn't just disappear into the ether. They could be rerun and resold and syndicated, and Hollywood could get a chance at the new brass ring, too.

The big American corporate sponsors and advertisers really liked the idea of eliminating live broadcasting and making programs on film. The programs could be made more glamorous, better suited to the ads, and there would be greater control over what was going out over the airwaves. The big sponsors could see the immediate success of filmed formula series like *I Love Lucy* and *Dragnet* – both of which had been made on film precisely for the purpose of syndicated reruns. As backers of a medium meant for selling products, advertisers and sponsors quickly recognized that filmed product was exactly what they wanted for American TV. Hollywood was quite willing to oblige.

By the mid-1950s, the U.S. TV networks (bowing to the wishes of the sponsors) had largely phased themselves out of in-house production (aside from their news shows) and were instead filling their broadcast schedules with filmed programs made by the big Hollywood studios, as well as by independent production companies which sprang up to take advantage of this significant shift in broadcasting practice: from live production to filmed production, and from in-house network-made programming to reliance on series made by production companies independent of the networks.

At the same time, the networks turned away from the practice of selling single sponsorship of a show, and began the now-standard practice of selling small spots of time on a program to lots of advertisers. With the increasing desirability of TV as an advertising medium, the networks steadily escalated their rates until in 1986 one thirty-second prime-time spot sold for an average of $118,400.

This increased ad-revenue for the networks was accompanied by a brilliant financing system which the networks arranged in the mid-1950s with the studios and companies that made the TV shows. A program-maker such as MTM, Disney, Lorimar, or MCA-Universal does not literally sell a show to a U.S. network. Instead, the production company rents the show or series to the network for a "licence fee" that covers (usually) two showings of the series by the network: first in prime time, and then as a rerun. Typically, this fee amounts to between 50 and 70 per cent of the total costs of making the program. This arrangement between a network and a production company is called "deficit financing". It is the structural key to the

American TV industry as a business, and it is also the key to television practice world-wide.

The shift to deficit financing in the mid-1950s completely altered the
economic structure of the industry and accounts for virtually all practices
that have continued since that time. Let's say it costs a producing company
$1 million to make each episode of an hour-long crime series like *Miami
Vice* – a production cost figure that is now typical for such American
shows. If the series runs for twenty-two episodes in a season, this means a
cost of $22 million to the producing company. The network, however,
licenses the series for, at most, 70 per cent of the costs, or about $15.4 mil-
lion – meanwhile selling ad-time on each episode at a rate that easily brings
in double the amount paid out for the series in licensing fees.

If the show becomes a hit according to the ratings, it is the network that
profits from the success, at least immediately, because it can then increase
its ad-rates for each episode throughout the season, without having to pay
out any more money to the company that produced the series. In fact, that
company would seem to be losing several million dollars on the deal
because of the deficit financing arrangement. The idea, however, is that the
production company will keep churning out episodes until it has, say, five
seasons' worth of product in-the-can. The company is playing a waiting-
game until its hit series (which has been making big bucks for the network
in ad-revenues) has had its two licensed network plays (a first run and a
rerun). Then the series can become a syndicated re-rerun outside prime
time: bought up by hundreds of local U.S. stations and "stripped in" to
their daily schedules – say, every day at 4:00 p.m. For the production company, the syndication of a one hundred-episode package to local broadcasters across the United States can easily bring in cumulative profits of
some $20 million: the prize for hanging in through the period of deficit
financing.

This change in the economic structure of TV production accounts for
many aspects of U.S. broadcasting. The "necessity" of building up an in-
the-can series package for syndication leads to an assembly-line approach
to production, complete with the product research and testing that goes
along with any mass-produced product. The "stripping-in" of syndicated
reruns by local stations (the cheapest way of filling their broadcast time)
means that usually the only thing "local" about a local station is its
weather reports.

Historically, each of these shifts – from live broadcast to filmed production, from in-house network-made programming to reliance on product
from studios and companies independent of the networks, from live drama
to filmed episodic series, and from in-house network financing to the

deficit financing arrangement – significantly altered the future of television. Combined, all these changes in the mid-1950s congealed the medium into a machine that was extraordinarily efficient on several fronts.

Perhaps the most significant efficiency was in the guarantee that henceforth programming would be error-free. The shift to film meant that television need only adopt the long-standing conventions of Hollywood film production in which no human mistakes remain as evidence to reveal the human and technological process. Live TV broadcasting had been constantly riddled with errors: delayed scene changes, actors missing their chalk marks to be left in shadow, cameras or lights blowing out, actors flubbing their lines or going "cold" in the middle of the broadcast, toppling sets, missed cues, a "corpse" accidentally shown crawling off the set. Gary Moore once went into a wild dance with his pants unzipped. Red Skelton, giving a sales pitch for pies made from Pet Milk, led a cow out in front of the camera and the beast promptly delivered its own form of editoral comment smack-dab in the middle of the commercial. On a CBC-TV variety show, the host was supposed to enter the set through a doorway flanked by ivy-covered trellises, but the door wouldn't open. The host panicked and tore his way through the trellis, catching his toupee in the ivy. On a Saturday morning kids' show, the host was overheard to say, just before the insert of a commercial: "I hope that keeps the little bastards happy."

So live television tended to continually expose its spectacle as a human construction. The shift to filmed product meant that television would be free of such errors, and it could maintain the necessary illusion of an unconstructed, seamless technological dream unfolding for audience pleasure. It would appear "perfect", or as "perfect" as possible, thereby enhancing the advertised products which also promised perfection and greater efficiency if only they were purchased.

One aspect of TV programming that, in a way, summarizes this whole ethos is the use of what's called "the sweetening machine" – the apparatus that generates prerecorded laff-tracks and applause-tracks to augment, or "sweeten" the sound-track of television productions. Not surprisingly, the sweetening machine was invented at the time the Hollywood studios and production companies were gearing up to take over from the networks the making of most product for the airwaves. A man named Charlie Douglass, who had been a sound technician at CBS, put together a machine that could reproduce a wide variety of laughs and audience responses – everything from a few quiet chuckles to uproarious crowds guffawing and applauding wildly.[1]

It was, of course, a timely invention in that most of the filmed shows would have no studio audience, especially the filmed sitcoms of the 1950s. Here, with Charlie's magic box, the production could be given the ambience of live TV. Better yet, there would be perfect control of this "audience". It would laugh at exactly the right moments, and to just the right degree. All you'd have to do is let Charlie orchestrate the giggles, whoops, groans, and bursts of hilarity and applause into a perfectly tuned and timed sound-track, and the production would be perfection epitomized, the greatest crowd-pleaser that ever hit the airwaves. Charlie Douglass first approached Desilu Productions in the early 1950s, and the rest, as they say, is history.

Lucille Ball and Desi Arnez had hired retired cameraman Karl Freund to design a complicated new system for filming simultaneously with three 35mm movie cameras, then editing these varying camera angles into a fully sound-synchronized program. But the resulting product lacked that ineffable quality of a live audience enjoying the comedy. Charlie Douglass's timely sweetening machine was exactly the means for providing a simulacrum of the missing ingredient.

Despite the fact that Charlie kept his machine shrouded in secrecy and under lock and key, other independent laff-men soon sprang up to rival his position in the production-perfect industry. But there was plenty of work for all. Canada, too, had slavishly followed U.S. production style, even though TV broadcasting here had been ostensibly established for quite different purposes and had no real reason to imitate. As filmed and then videotaped production overtook live broadcasting in Canada, Canadian TV producers also wanted to sweeten studio-audience response, or replace it where necessary. Peter Campbell, a member of CBC's sound-effects department, recalls, "Until recently, guys like John Pratt and Charlie Douglass from the States would come to Canada and sweeten the shows."[2]

So sweetening was becoming not just another level of expertise, but an international product nicely standardized and sold across the border. Eventually a Canadian, Rafael Markowitz (who later moved to California), entered the busy sweetening scene with his own machine. Says Campbell, who worked with Markowitz for years, "It took us two-and-a-half years to get what we wanted" in the necessary variety of laffs and applause. After all, competition was fierce. Joe Partington, producer and co-creator of CBC's *Hangin' In*, recalls, "We used to let Markowitz record the live audience for *King of Kensington* in order to add to his repertoire, and we always wondered if Canadian laffs were being used for American shows."[3] In any case, the Canadian TV producers were strictly dependent on these traveling laff-men who cross the border with ease. Says Partington, "When you rented [laffs] from outside people, it was a real secret how the machine worked." Douglass, Pratt, Markowitz, or whoever was working the sweetening machine would hide it under a table during operation so that CBC-TV personnel wouldn't be able to see how it worked.

But finally, CBC's crew of sound engineers took on the challenge and figured out how to build their own sweetening machine. Nobody I talked to could remember exactly what the historic date of its first use was, but everybody agreed that it was sometime in the late 1970s, during the making of *King of Kensington*. According to Partington, most of the laffs and applause were accumulated from studio-audiences for *Wayne and Shuster* and *King of Kensington* – that "they are exclusively Canadian laffs". Of course, the repertoire of the sweetening machine is being continually upgraded, adding diversity and nuance to its collection. Says Campbell, "You want to get a special feel, create character to match the situation." That means having a wide range of response-simulacra from a wide range of different-sized audiences. "The machine can respond quite sensitively," says Partington, "generating little touches like 'ooohs and aaahs'."

For a TV show like *Hangin' In,* taped without a studio audience, the sweetening technician and the associate producer go through the videotape during an audio mix and decide precisely where and what kinds of sweetener are needed. The machine has twenty-four tracks and can hold six of the prerecorded laff / applause cassettes at once, making possible a carefully nuanced sound that is different for every show. For shows that do have studio audiences during the filming or taping of the production, the same process is used. They didn't particularly like the jokes? No sweat. At the audio mix just cut out their lame response and put in some sweetener. Was the applause at the end a little weak? No problem. The repertoire of the sweetening machine can match the dimensions of the room, the size of the audience, and put into the final sound-track the degree of applause and appreciation that *should have happened* for the show.

Not surprisingly, the sweetening machine has been used in live broadcasts as well. U.S. network people slip a little sweetener into the Rose Bowl Parade to get that little ripple of applause that should happen as each float goes by.[4] A sweetening technician is always on hand at live broadcasts of telethons and awards ceremonies, including the Academy Awards, to spice up the production. An alert technician can save face for anybody who cracks a dumb joke, giving them a small dose of laffs so that the home audience, at least, doesn't think they're total jerks. But if sweetener is used for parades, telethons, and awards ceremonies, could it also be used for telecasts of political conventions, press conferences, or public speeches by politicians? "Not to my knowledge," says Peter Campbell, "but that doesn't mean it isn't done. Just that I've never heard of it."

With all of its illusion-making production conventions that guaranteed tight control and "perfection," the technological cataract was given a rose-coloured tint more suitable to the role of spectacle itself. As Guy Debord writes in his extraordinary book, *Society Of The Spectacle*: "The spectacle is the nightmare of imprisoned modern society which ultimately expresses nothing more than its desire to sleep. The spectacle is the guardian of sleep."[5]

Underneath the rose-coloured tint is a nightmare, the nightmare of imprisoned modern society with its "anti-people climate", its expanding war-machine, its technological simulacra replacing human endeavour and involvement, its machine-like values and characteristics overtaking all others. As spectacle, television mirrors this nightmare, even is this nightmare, at the same time that it must guard the sleep of its mass audience. Both contributing to and protecting the "desire to sleep" (that is, remain unconscious), television spectacle needs its rose-coloured tint, its hue of beneficent containment around harsh, messy reality, in order to perpetuate the larger nightmare unfolding beyond its screen.

As TV spectacle has gotten "better" – more "perfect", more pleasurable, more involving, more "realistic" – reality itself has gotten worse. This is obvious not just in the wholesale pollution of the natural environment, the deterioration of community and cities, and the gradual poisoning of everything necessary for human survival, but in the supremacy of technological expansion over all else. As reality worsens, we have increasingly turned to the screen, with all its controlled and sweetened perfection. Isolated in our homes, separated from one another by this very screen, we fulfil our assigned role of consumption and spectatorship while the dictates of a larger machine, far less "sweet", proceed unimpeded.

Part II

The Guardian of Sleep

5

Fine-Tuning:
TV, Desire, and the Brain

We are not concerned with getting things across to people as much as out of people. Electronic media are particularly effective tools in this regard because they provide us with direct access to people's minds.

TONY SCHWARTZ, *The Responsive Chord*

FLASHBACK 1977: I am sitting in the office of Tony Schwartz, adman, protégé of Marshall McLuhan and "electronic guru" in his own right. Out of his New York City studio Schwartz has masterminded hundreds of advertising campaigns, not only for commercial products but also for many politicians, including two U.S. presidents. It is just a few months after the election of Jimmy Carter, and Schwartz is being credited with dramatically turning the campaign around after the candidate had slipped thirty points in the polls. Schwartz was hired at that crucial juncture by Carter's top media advisor, Gerald Rafshoon, to come in and design TV ads that would get the Carter Campaign back on track. Obviously, Tony Schwartz had worked his media magic. Carter had subsequently been sworn in to the highest office in the land.

"What were Carter's media advisors doing wrong before you were brought in on the campaign?" I ask.

"I felt that they were using television as a window on the campaign that was taking place out around the country," answers Schwartz. "They were photographing meetings that they held where Carter was speaking to

67

people, and they were presenting his thoughts from those meetings where he was speaking, facing other people. They were editing those speeches and bringing them into your home. So when you were looking at Carter on TV, you were actually looking at him facing to the side, or looking someone else in the face. You had no eye contact with him. His projection-distance was not to you. You were watching him campaign to other people. Now, I feel that since he's coming into your home on this medium, he may as well campaign to you."

"Instead of having TV be a window on the world of the campaign, how did you use it?"

"I was using it as a door to your home, even as a door to your mind," says Schwartz. "My idea was to utilize media to attach Carter to the concerns of the American public."

Jimmy Carter stands on the Concord Bridge, a scene resonant with historical nuance and patriotic sentiment, as well as embodying the symbolism of transformative change: a bridge. At a signal from the director, Carter looks directly into the camera, his eyes calm and sincere. "We've lost our vision of what this country ought to be," he says, his voice soft and pious, with just a touch of sadness. "We've lost our ability to work together. What we need is a government as good as our people." Once he is settled into the White House, Jimmy Carter begins to study Tony Schwartz's book, *The Responsive Chord*.[1]

□

Flashback 1979: Focus groups conducted by Ronald Reagan's media advisors have uncovered a serious problem with their candidate's image. Using free-association testing and probing questions, teams of psychologists have pinpointed a widespread fear among the populace that candidate Reagan, if elected, could get the United States into a nuclear war.

"That's the biggest liability we have with the Governor's image," says media aide Stuart Spencer before the election.[2] "He is one of the great communicators of his generation," says Peter Dailey Sr., Reagan's media director. "Our only problem is how to get that warmth compressed into thirty seconds of television."[3]

Candidate Reagan sits in a dark leather chair in the "oval office" set. He wears a presidential deep-blue suit. On cue he looks directly into the camera, his face grandfatherly and kind. He speaks. "Nancy and I have travelled this great land of ours many times over the years and we've found that Americans everywhere yearn for peace just as we do." His voice is soothing and sincere. "It is impossible to capture in words the feelings we have about peace in the world, and how desperately we want it for our four children and our children's children."

The compressed warmth works, banishing the fear of nuclear war associated with candidate Reagan, who soon after moves in to the real Oval Office. "I can't imagine," says political consultant David Sawyer, "how you could win a modern election without the help of attitudinal polling."[4] Or without the help of television, which has become the primary means for transforming polling results into the redesigned image.

□

"Recent attitude-change," wrote Schwartz in 1973, "has shown that the most favorable condition for affecting someone's attitude involves a source the listener depends on or believes in, and yet one he does not actively or critically attend."[5] The word "listener" in this statement makes us think of radio, but it is, of course, television that has now become the "source" everyone depends on and believes in.

The fact that TV is a source not actively or critically attended to was made dramatically evident in the late 1960s by an experiment that rocked the world of political and product advertising and forever changed the ways in which the television medium would be used. The results of the experiment still reverberate through the industry long after its somewhat primitive methods have been perfected.

In November 1969, a researcher named Herbert Krugman, who later became manager of public-opinion research at General Electric headquarters in Connecticut, decided to try to discover what goes on physiologically in the brain of a person watching TV. He elicited the co-operation of a twenty-two-year-old secretary and taped a single electrode to the back of her head. The wire from this electrode connected to a Grass Model 7 Polygraph, which in turn interfaced with a Honeywell 7600 computer and a CAT 400B computer.[6]

Flicking on the TV, Krugman began monitoring the brain-waves of the subject. What he found through repeated trials was that within about thirty seconds, the brain-waves switched from predominantly beta waves, indicating alert and conscious attention, to predominantly alpha waves, indicating an unfocused, receptive lack of attention: the state of aimless fantasy and daydreaming below the threshold of consciousness. When Krugman's subject turned to reading through a magazine, beta waves reappeared, indicating that conscious and alert attentiveness had replaced the daydreaming state.

What surprised Krugman, who had set out to test some McLuhanesque hypotheses about the nature of TV-viewing, was how rapidly the alpha-state emerged. Further research revealed that the brain's left hemisphere, which processes information logically and analytically, tunes out while the person is watching TV. This tuning-out allows the right hemisphere of the brain, which processes information emotionally and noncritically, to function unimpeded. "It appears," wrote Krugman in a report of his findings, "that the mode of response to television is more or less constant and very different from the response to print. That is, the basic electrical response of the brain is clearly to the medium and not to content difference.... [Television is] a communication medium that effortlessly transmits huge quantities of information not thought about at the time of exposure."[7]

Soon, dozens of agencies were engaged in their own research into the television-brain phenomenon and its implications. The findings led to a complete overhaul in the theories, techniques, and practices that had structured the advertising industry and, to an extent, the entire television industry. The key phrase in Krugman's findings was that TV transmits "information not thought about at the time of exposure."

In the human brain, the neo-cortex (the locus of human intelligence) is divided into left and right hemispheres, which are interconnected by a central bundle of nerve pathways called the corpus callosum. The left hemisphere, which controls and is also affected by the right side of the body, tends to operate according to sequential logic: rationally analyzing discrete parts to arrive at a logical conclusion. The mode of approach of the right hemisphere, which controls and is also affected by the left side of the body, is to recognize instantly, or all at once, patterns, spatial relationships, and whole gestalts, without the part-by-part analysis typical of the left hemispheric mode.

The right hemisphere excels at tasks like the recognition of faces and places, the completion of incomplete patterns (as when we see a rough caricature, but recognize the person depicted), and the response to archetypal or symbolic aspects of imagery. While the left hemisphere seems to have a coldly analytical style of processing information, the right hemisphere is emotionally involved, perceiving the world in terms of past emotional experiences whose moods, sensations, and images have left a memory trace, sometimes not consciously remembered. The individual's worldview, or basic feeling-orientation in the world, seems to be consolidated in the right hemisphere, which functions according to imagery, analogy, feeling-states, moods, and sensations.

In the healthy adult, both these modes of operation combine forces to fully process information from the environment. Through the nerve pathways of the corpus callosum, the right and left hemispheres pool their relative strengths, moment by moment, to yield a balanced response to a situation. The right hemisphere, which is less verbally oriented than the left, needs the left hemisphere to articulate what it knows. Similarly, the left hemisphere needs the right one to provide contextual information to round out its sequential and logical analysis. Thus, critical assessments and logic are tempered and informed by the emotional and intuitive information that the right hemisphere provides. And emotional-intuitive insights or hunches are tempered by critical logic and analysis.

To live sanely in the world, one needs both hemispheres of the neocortex fully engaged in one's experience, though there are times and situations when one or the other hemisphere is dominant. But what Krugman and other brain-researchers have found is that watching television tends to shut down the left hemisphere, disengaging the information processing of this area of the brain. The appearance of alpha rhythm in the brain-waves indicates this shutting down. To fully understand what Krugman meant when he wrote that television transmits content "not thought about at the time of exposure", we must recognize that he is referring to logical-critical thought – the thinking of the left hemisphere. The right hemisphere, which has its own mode of thinking, stays tuned.

The television transmission process is unlike any other medium. There is no actual picture being projected, as there is from a movie projector. If you pick up a piece of motion picture film, you will see that each frame on it is a visible photograph, which runs through a projector at twenty-four frames per second. But if you pick up a piece of videotape you will see no visible image on it. Like audiotape, it holds nothing but electrical impulses. And in live transmission, there need not even be the intermediary record of tape holding the electrical impulses. So with television, there is no actual image to speak of.

Instead, television works by electronic scanning. Tiny dots of light on the screen are lit up, one at a time, by the firing of cathode ray guns across alternate lines on the front of the picture tube. The succession of glowing dots moves rapidly across and down, along alternate lines, in a "sweep" that lights up the first series of phosphors. In all there are 525 such lines of miniscule dots on North American TV sets. During each one-thirtieth of a second, the scanning process completes two full sweeps of the screen, once on each alternate set of lines, to create, by electrical impulse, the whole mosaic of an instant's image. In terms of micro-seconds, however, there is actually never more than a single dot of light glowing on the screen. Our eyes receive each dot of light, sending its impulse to the brain. The brain records this bit of information, recalls previous impulses, and expects future ones. We "see" an entire image because the brain fills in or completes 99.999 per cent of the scanned pattern each fraction of a second, below our conscious awareness. The only picture that ever exists is the one we complete in our brains.

"With film," says Tony Schwartz, "the picture is there, projected for short bursts of time, changing twenty-four times a second. Your brain fills in the motion between frames. But you take a giant step into another area when you go to television in that the picture's never there. You're both filling in the motion *and* the picture. You're very involved, very busy."[8] But this involvement, this busyness, is apparently only on one side of the neocortex. As the studies have found, the left hemisphere tunes out.

Researchers Fred and Merrelyn Emery at the Australian National University suggest that the scanning process of television is ideally suited to the right hemisphere's mode of processing information – through the completion and recognition of patterns.[9] Though the scanning process actually delivers a series of glowing dots in two sweeps, the electronic speed of transmission allows them to be made sense of as a mosaic, filled in by the right hemisphere as recognizable images. The Emerys also speculate that the left hemisphere tunes out because it quickly becomes habituated to the scanning dot and, lacking any need to respond critically or analytically to it, slows down and goes into alpha. There is also the possibility that the left hemisphere becomes overloaded by attempting to attend to scanning at electronic speeds. Such overloading permits direct access to the right hemisphere, where the viewer is more noncritical and suggestive. The literature on brain-hemisphere research reveals that without the critical left hemisphere analyzing and involved, the right hemisphere is free to accept and act upon suggestions or commands, even nonsensical ones.

The right hemisphere of the neo-cortex is often scientifically referred to as the "nondominant" hemisphere. This might remind us of the fact that, in a patriarchal society, the processing skills and contextual information provided by the right hemisphere have been consistently devalued. Prizing logic, analysis, and strictly rational thinking (with its resulting numerical and digital emphasis), the patriarchy allows such thinking to dominate over the seemingly less rational skills of pattern recognition, grasping a whole gestalt (intuition), recognizing the feeling-values of images or situations, or being attentive to unquantifiable aspects. Precisely because these skills have been devalued, they can be more readily exploited, and are. In recognizing that the TV / brain phenomenon leaves the nondominant hemisphere open to suggestion, the industry has rushed to find the best ways of utilizing that situation.

There is another significant finding of brain hemisphere research, which has to do with selecting the particular mode of thinking by which we approach any activity. As Robert Ornstein, leading researcher at the Langley-Porter Neuropsychiatric Institute in California, puts it, the hemispheres are apparently specialized for the kind of thought or information-processing people choose to use, not necessarily for the type of material they confront.[10] It is possible for TV viewers to watch in the analytical-critical frame of mind: noticing the use of specific camera angles, camera distances, editing, sound-image relationships, or framing and composition, and relating such techniques to the creation of overall meaning and affect. In a sense, this entails conscious resistance to what Krugman identified as "the basic electrical response of the brain to the medium". But the typical viewing situation discourages such conscious attentiveness. Most of us watch TV when we are already tired from a day's work. We are not intent on focused analysis, nor do many people think of the medium as one that warrants critical attentiveness. Moreover, as McLuhan recognized, the response to the medium may well be at the level of the nervous system, working its larger effect regardless of critical analysis on the part of the viewer.[11]

□

By 1971, as the TV / brain phenomenon began to be known throughout the communications world, ad agencies started conducting their own brain-hemisphere research and reassessing their advertising techniques in the light of new knowledge. One of the first things to be toppled was the old model of communications: sometimes called the hypodermic – or transportation – model of communications.

Traditionally, there are four basic elements in any act of communication: a *sender* of a *message* uses some *medium* to reach a *receiver* of that message. The old hypodermic / transportation model arranges those four elements as isolates in a linear sequence. The elements all converge on the human receiver at the end of the line. As its name suggests, the model posits the act of communication as a process of injecting or transporting some message to the receiver, who seems to be waiting passively for its arrival. Using this model, ads were designed to spell out the logical advantages of buying a certain product, and market researchers tested viewer recall: determining the number of claims and qualities mentioned in the ad, information that viewers were able to remember after seeing it. Like the model itself, the testing procedures were based on the delivery of a clear-cut package through the process – as in mail delivery. How much of the message got through to receivers, as evidenced by their "learned recall" of product-claims and logical reasons for buying, indicated the success of the ad.

But by the late 1960s and early 1970s the marketing industry began to realize that this communications model did not seem to fit the workings of the electronic media. Even more significantly, the industry realized that its logical appeals for buying a product were directed mainly at the brain's left hemisphere – that side of the brain which appeared to tune out during TV viewing. In 1971, NW Ayers / ABH International, one of the largest ad agencies in the world, began using EEGs to evaluate its TV commercials for several clients, including AT&T. The agency's subsequent ads for Bell

Telephone ("Reach Out And Touch Someone") are now considered classics of applied brain hemisphere research.

What happened, they found, was that the old hypodermic / transportation model oversimplified the communications process, giving too much emphasis to the "message" while downplaying the importance of the other elements. It also tended to isolate these elements arbitrarily. (As McLuhan and then Krugman were suggesting, the *medium* is the message.) But perhaps more importantly, the old communications model implied that the human receiver was a kind of tabula rasa, a "clean slate" or empty vessel, being passively injected or filled for good or ill by the media.

According to the new model of communications emerging out of the brain hemisphere research, the human receiver is by no means an empty vessel waiting to be filled with a potent message. Rather, the human receiver is a bundle of needs (many of them unconscious or below the threshold of awareness) and a compendium of emotional experiences (many of which are common to all of us as members of this society). As well, that receiver lives in a highly "coded" society, usually without being explicitly conscious of those codes. For example, most of us have certain notions about how a "credible politician" should dress, talk, look, behave, live. Somebody campaigning in a T-shirt and blue jeans would, for many people, be breaking one of the dominant codes at work in the situation. Usually we don't recognize the codes in our society until somebody calls them to our attention, often simply by transgressing them.

With the new communications model, then, the receiver is a highly involved participant in the communication. The goal is to shape the message so that it matches the unconscious needs, emotional experiences, and coded expectations of the desired audience – so that it speaks to, or resonates with, their deepest feelings and beliefs. This isn't putting something into the receivers, it's drawing something out of them and attaching it to, or labelling that emotion with, the product being advertised.

The Bell Telephone ads of the early 1970s fully lived up to this newly emerging model. To see how the "Reach Out And Touch Someone" ads work, we need only recognize that most adults have moved away from the family home, are pretty alienated from the old family ties, and are vaguely guilty or sentimental about distant relatives. In other words, we have some highly coded notions about what a "good family" should be like in our society. But at the same time our actual lives don't particularly live up to those codes. All these nebulous feelings are already in us. The ads need only resonate with those feelings, through music and images of caring families, and label the feelings for us – as the "long-distance feeling". The ads make no logical claims whatsoever, nor do they even seem to be selling anything. They merely provide an experience of images and sounds that are pleasant and familial.

According to the new model of communications, the most effective and affective TV ads work just this way: they resonate with our deeper feelings, beliefs, and coded expectations – touching a place inside us and triggering an emotion – which can then be labelled for us as product x or candidate y.

With his book *The Responsive Chord,* published in 1973, Tony Schwartz gave a name to the newly emerging communications model with his resonance theory of the electronic media. According to this theory, an ad is effective if it "strikes a responsive chord" or resonates with the deep feelings of the receiver. "Certain stimuli, in the proper context," wrote Schwartz, "can recall experiences that we could never remember at will. I do not care what number of people remember or get the message. I am concerned with how people are affected by the stimuli."[12] The key difference is between learned-recall and evoked-recall.

"I can show you this difference," says Schwartz as we sit in his office, "by saying do you remember how I said I used Carter on television? Now you'd have to go back to what I said and try to recall that. Maybe you learned it and maybe you didn't. That's different from what I'm going to do now. I'm going to get you to say some words that I've never told you, but that I know are inside you. For instance, things go better with ..."

"Coke," I say.

"For the rest of your ..."

"Life," I answer.

"You're recalling these connections," says Schwartz, "and I've never told you those. We've experienced them in our life. Now I'm utilizing a different type of recall when I use that. It's the type of recall that comes from associations. It's evoked-recall. Association is evoked connection."

I hesitate to blurt out to Schwartz that since he designed that Coke ad, he did, in effect, tell me those words. But I know what he's getting at. None of us "learned" those associative strings of words that conjure up the right response – at least not in the sense of consciously memorizing them. Yet we retain them, they are in our long-term memory bank to be evoked by association. Associative memory, the locus of evoked-recall, is part of the right hemisphere of the brain.

With the resonance model of communication, the advertising industry began to become fully in tune with the TV / brain findings. The right hemisphere, with its emotional and not fully conscious world-image, its propensity for processing imagery and completing patterns, its storehouse of associations, connections, and long-term memories and feelings below the threshold of awareness, and its receptiveness to the television medium without the critical, logical left hemisphere to interfere, was precisely the place to strike responsive chords. The new resonance model of communications acknowledges an element that the traditional model didn't consider significant. Along with the sender of a message using a medium to reach a receiver, the new model recognized that the situation in which this process takes place is highly important.

Here again, we have to recognize a distinct difference between movie-going and TV-watching. Going to the movies has always retained some vestiges of the conscious act. Its context reminds us of that. We make our selection, leave our homes, travel to the site of the screening, and enter a public space where a monetary transaction underlines the eventfulness of what we are doing. All these aspects tend to emphasize the ritualistic nature of movie-going – in the sense that we consciously decide to enter the space of ritual where we participate for a time in a communal experience.

Of course, as a ritual movie-going is not as fully communal as live performance, or even live participation as subject, not spectator. Nevertheless, going to the movies retains some of the qualities of traditional ritual: in which we consciously agree to enter numinous space and temporarily relinquish our self to a communal experience that is transformative. Arguably, it is the degree to which we are conscious of entering the ritualistic space of the archetypes that allows us to integrate the transformative experience, or to critically investigate its appeal.

By contrast, television-watching is unconscious, in part because of the situation. It is "on" in our private space – the space not of public ritual, but of personal relaxation. If initially a conscious selection of a program has motivated us to turn the TV on, the flow of programming may gradually change the nature of the act to that of simply watching TV. The markers that make movie-going a conscious ritual are absent from the television situation. Its immediate accessibility and twenty-four-hour availability make it entirely uneventful, familiar, intimate.

Thus, even for the media-illiterate, the act of movie-going allows for boundaries around the event – boundaries that at the most minimal level allow for a certain separation between the time / space of the movie, and the time / space of actual life. Television, however, tends to be completely without boundaries, overflowing. Though it is entirely communal and public in the sense of its vast audience, it pretends to be for your eyes only, as personal as your mirror. Without being fully conscious of doing so, we step into its ritualistic space: open to archetypal transformations without even knowing it.

□

"We are the descendants of Freud and Jung," writes Jungian analyst and author Marion Woodman, "and while poets and madmen had free access to their unconscious before those two giants, the world of the archetype is now an open market for the general populace without any ritual containment."[13] As watching televison replaces all other social rituals, it becomes the unacknowledged ritual container: triggering archetypal responses and naming them with the brand names, consumer mores, celebrity faces, and catch-phrases of the marketplace. These TV mediations become what we have in common.

Throughout the 1970s the industry began testing to see just what stimuli would be most effective in right-hemisphere appeal. Following the lead of General Electric and NW Ayers / ABH International, marketing organizations such as the Simmons Market Research Bureau, Cockfield, Brown & Company Ltd. in Toronto, and the New York ad agency KSW & G Inc. launched their own TV / brain research in order to better serve their clients. Hiring researchers such as Sidney Weinstein (a neurophysiologist in Danbury, Conn.) and Richard Davidson (director of the Laboratory for Cognitive Psychobiology in New York), such agencies successfully pinpointed the stimuli that would best serve their purposes.

The laboratory research revealed that the right hemisphere is engaged by tone of voice, rhythm and melody, rhyme and harmony, and pictorial emotional triggers. It is involved in evoked-recall, rather than learned-recall, and engages in free-associations, and unconscious connections. Although the right hemisphere had long been thought of as the nonverbal hemisphere, researchers found that it does have its own word-generating skills. Its speech is comprised of expletives, emotional words, clichés, and buzzwords, and the evoked associative response that Schwartz had demonstrated with me.

As a result of their findings, television advertisers changed tactics. The singing commercial, the use of jingles, and the short carefully worded slogan became predominant techniques used in virtually every TV ad. The rhythm of the phrasing, or the tune of the melody with words, are retained far longer than rhetorical claims, but without being consciously learned. Some jingles take advantage of the right hemisphere's subjective bias. Ginny Redington's jingle for McDonald's during the 1970s ("You, You're The One") inspired later appeals by other agencies: "You, You Never Looked So Good", "You Asked For It, You Got It!".

TV ads are one of the few forms of television programming that address us directly, both through the direct gaze of the speaker and verbal content. This is a powerful technique for right-hemisphere resonance since the gaze of the "other" is always an emotional experience, reaching deep layers within. Lacanian theory is relevant here, raising issues of the constitution of the self through the experiencing of the other's gaze. I have little doubt that the teams of psychiatrists and psychologists working for the marketing field are fully aware of the resonant power of the direct gaze and use it quite skilfully and purposely in TV ads. It was, for example, the change in technique that Schwartz introduced into the Carter presidential campaign: from profile shots to direct address to the home viewer.

Following on Bell Telephone's "long-distance feeling" of the early 1970s, it became commonplace for ads to label our feelings for us. So we're told that we're "feeling 7-Up", and "Oh What a Feeling – Toyota". Or an ad will incorporate a popular song from the past and establish an association between the pleasant memories it evokes and the image of the product that appears with it. Steve Karmen, "king of the jingles", pioneered the technique of the incomplete lyric ("You can take Salem out of the country, but ..."), which relies for its effectiveness on the right hemisphere's propensity for completing incomplete patterns. His jingle for General Tires during the 1970s ("Sooner Or Later You'll Own Generals") involved the power of a command – which the right hemisphere accepts in its noncritical mode – couched in a pleasant melody. Karmen's four-note jingle for the "I Love New York" campaign is remarkably memorable and obviously emotive. The 1970s ads for the Ontario Milk Marketing Board, showing calm and lyrical settings and people with the accompanying theme song "Thank You Very Much, Milk", resonated with memories of mother and infantile feeding. The sunlit, hazy imagery, usually in slow motion, coupled with the nursery rhyme qualities of the jingle, evoked pleasant child-like experiences that could be attached to, or labelled with, the carton of generic milk on screen.

Often TV ads incorporate a nonlinear series of pleasant or fascinating images followed by a simple captioning device: either a shot of the product, or a slogan, or both. The 1980s Pepsi semiphore ad – with its woman flag-waver and its spurting cans – makes no linear sense but is engaging visually. This fascination is labelled with a seeming non-sequiter: "New Diet Pepsi – Taste Above All". The old plot-based ads, where a thirty-second mini-drama unfolded, are becoming increasingly rare, as advertising abandons linear, left-hemisphere forms that are not as involving in repeated viewings as the more purely visual / emotive styles. Like rock videos, nonlinear ads are the forms of programming most suited to the right hemisphere.

With the right hemisphere resonating to the language of the unconscious, this archetypal content is at the disposal of the industry. Recently, advertising has exploited the god-archetype: our deep sense of a greater spirit that transcends individual existence. Since the mid-1970s the use of the word "spirit" has been common in ads, which make subconscious links between mundane products like cola ("Pepsi Spirit"), a department store ("Simpson's Spirit"), or a hotel ("Spirit of Hyatt") and our sense of, and associations with, the Godhead. Similarly, the word "light" is sometimes used to trigger associations with divinity and transcendent power.

But media involvement with the deity also goes deeper than verbal references. Tony Schwartz, for one, is well aware of the similarities between the electronic media and the traditional concept of God. His 1983 book, *Media: The Second God,* is based on this correspondence wherein disincarnate media images and voices (omnipresent throughout the environment and seemingly omniscient) are consumed and internalized by us and direct our volition like promptings from a deity.[14]

In *The Origin Of Consciousness In The Breakdown Of The Bicameral Mind,* psychologist Julian Jaynes also poses an interesting theory, one based on archaeological research into ancient cultures, and on 1970s brain hemisphere research. Jaynes speculates that modern consciousness as we know it did not exist before the first millenium B.C. Though he never uses the term "matriarchy", it is clear that Jaynes is referring primarily to those ancient cultures in his discussion of what he calls "the bicameral mind". That mind, according to Jaynes, was characterized by a oneness with the cosmos and a lack of a personal ego separate from the will of the gods because, he implies, the left hemisphere of the brain had not become specialized and distinct from the holistic right hemisphere. He writes of the ancient cultures as being based on directives from their gods, or what he calls their "divine hallucinations". He speculates that these messages from the gods must have originated in what we would now call the right hemisphere of the brain:

> The gods were in no sense "figments of the imagination" of anyone. They *were* man's volition. They occupied his nervous system, probably his right hemisphere, and from stores of admonitory and preceptive experience, transmuted this experience into articulated speech which then "told" the man what to do. That such internally heard speech often needed to be primed with the props of the dead corpse of a chieftain or the gilded body of a jewel-eyed statue in its holy house, of that I have really said nothing. It too requires an explanation.[15]

Jaynes argues that with the separation and specialization of the two brain hemispheres (that is, "the breakdown of the bicameral mind"), modern consciousness arose: providing a sense of individual self-volition triumphing over "divine" commands arising from the right hemisphere. Rational, analytical thought – and the sense of a personal ego separate from the will of the gods – was thus a significant breakthrough for modern consciousness: giving rise to individual decision-making, responsibility, personal will.

What is of interest in Jaynes' text is not what it has to say about ancient cultures (his perspective is both patriarchal and modernist in its prejudices), but what it reveals about current brain hemisphere research and its implications. If any era is to be characterized as operating from a "bicameral mind", it may well be our own. With the TV / brain phenomenon and its exploitation by the television industry, we are in danger of collectively responding to the god-like promptings of media echoes in the right hemisphere as though they were our own volition. If, as the research indicates, the left hemisphere is effectively disengaged during TV-viewing unless we make the conscious effort to attend critically, then as a culture we are being swept into the totalitarian embrace of man-made media gods. We will be sacrificing personal consciousness to their "divine" promptings.

□

"Television is God," said Jerry Grafstein, media advisor to the national Liberal Party in Canada.[16] Addressing the Montreal-based advertising association Societal in March 1983, Grafstein focused his speech on the TV / brain phenomenon, outlining the political possibilities of the applied research. Like their American counterparts, Canadian political media advisors are applying to their own field the body of research used by product advertisers. Admen work for products and politicians alike. Canadian adman Jerry Goodis's clients have included Speedy Muffler, Wonder Bra, and the Liberal Party. Gerald Rafshoon has worked for Sears Roebuck and Jimmy Carter, among others. Tony Schwartz's clients have included Coca-Cola and Carter. Peter Daily Sr. took leave from his Los Angeles ad agency to be media director for Reagan's 1980 campaign, having previously managed to get Nixon elected in 1972. What they know about advertising in the commercial product field is applied to the political arena as well. The use of techniques derived from knowledge of right-hemispheric processing and appeals is now standard procedure in political campaigns and public relations efforts once in office.

Not surprisingly, television has come to play *the* predominant role in politics. During the 1980 presidential race, Jimmy Carter and Ronald Reagan both spent more than half of their total campaign funds on paid TV spots – the highest percentage that candidates had ever devoted to television up to that date. In the 1980 Canadian general election, sales of political party ads on TV networks were nearly double those of the previous campaign in 1979.

That was the year that U.S. political consultant Hal Evry said Canadian politicians were thirty years behind the times for wanting to actually appear in public, shake hands, and be seen in the flesh. Evry believes strongly in the effectiveness of the paid commercial TV spot and prefers his candidate-client to simply lay low during an election campaign. "You can't be misquoted if you don't say anything," he says.[17] Moreover, you can't reveal your human faults if you're not seen in reality. So thoroughly does Hal Evry believe in the perfectly scripted image that he has said he would even simulate campaign travel in a TV studio, complete with stage-set of a facsimile train interior, if a client would allow it. Evry claims a 94 per cent success rate for his political clients, who have included at least eight Canadian members of parliament.

This clear preference for the carefully orchestrated television image has behind it the awareness of the TV / brain phenomenon, wherein the left hemisphere, with its analytical-critical processing, easily tunes out. Moreover, since the majority of viewers now get all of their news from television, there is little critical information to balance the right hemisphere's impressions.

Political media advisors try to downplay the effectiveness of their TV spots. After speaking to a *Time* magazine reporter about the $18 million earmarked for Reagan's 1980 campaign ads, Peter Daily Sr. hastened to add that "People aren't really intently listening. TV is a way to turn yourself off."[18] He then went on to inform the reporter about the TV / brain hemisphere research, but conveyed the impression to the reporter that "the electronic brain scan of a person watching TV is remarkably similar to that of the same person sleeping soundly". Instead, as we know from the research, it is the critical hemisphere that turns off, while the dream-like right hemisphere stays tuned. Perhaps not surprisingly, careful observers of Reagan's televised speeches have noted that the word "dream" has been a recurring and favourite word-choice in his rhetoric: matching the assumed hemispheric state of his viewing audience.

During the 1970s, king of the jingles Steve Karmen was approached by Richard Nixon, Gerald Ford, and George Bush in turn. Each wanted him to work on their political ads. Karmen turned them all down, later explaining his refusal with the comment, "You wind up with President Toothpaste and Senator Cola".[19] Obviously, Ford, at least, found someone else. Political consultant David Sawyer says Gerald Ford's campaign surged back to a near victory in 1976 partly because of the Ford jingle: "I'm Feelin' Good About America / I'm Feelin' Good About Me."

"You can't advertise a politican as if he were Anacin," says Sawyer. "You know, the box and the name, the box and the name – all that repetition. People's feelings about politicians are too complex. But you can work with those feelings if you can find out what they are."[20] Psychographics and extensive polling techniques are the means for surfacing those feelings and designing media events to alter whatever image problems have been identified. "In a sense," says one Toronto adman, "what we're doing is wrapping up your emotions and selling them back to you."

"Party support is a mirage," says Hal Evry. "What has the real effect today is television, not party support. If you appear to be all things to all people, get on television and don't say anything but make it sound good, you can get three out of four to like what they read into you."[21] Thus, the television process triggers the mass projections of a populace that has been probed, polled, and psychoanalytically researched, so that its mass projections will serve the greater systems at work.

By the 1980s, the politics of those triggered mass projections were being switched from the individualism of the Me Decade – "You, You're The One" – to a recognizable neo-nationalism. Ginny Redington, jingle-writer for that 1970s McDonald's ad, explained in 1981 that the ad industry had moved to "the exact antithesis" of its former stance. "Nationalism is a success formula," she admitted. "Everything is America. 'Clean your face, America.' 'Brush your teeth, America.'"[22]

As Herbert Krugman noted in the research that transformed the industry, we do not consciously or rationally attend to the material resonating with our unconscious depths at the time of transmission. Later, however, when we encounter a store display, or a real-life situation like one in an ad, or a name on a ballot that conjures up our television experience of the candidate, a wealth of associations is triggered. Schwartz explains: "The function of a display in the store is to recall the consumer's experience of the product in the commercial.... You don't ask for a product: The product asks for you! That is, a person's recall of a commercial is evoked by the product itself, visible on a shelf or island display, interacting with the stored data in his brain."[23] Just as in Julian Jaynes's ancient cultures, where the internally heard speech of the gods was prompted by props like the corpse of a chieftain or a statue, so, too, our internalized media echoes are triggered by products, props, or situations in the environment.

As real-life experience is increasingly replaced by the mediated "experience" of television-viewing, it becomes easy for politicians and market-researchers of all sorts to rely on a base of mediated mass experience that can be evoked by appropriate triggers. The TV "world" becomes a self-fulfilling prophecy: the mass mind takes shape, its participants acting according to media-derived impulses and believing them to be their own personal volition arising out of their own desires and needs. In such a situation, whoever controls the screen controls the future, the past, and the present.

☐

"Here's something to forget," says the male narrating voice over the blank TV screen. "You'll watch it," the voice continues in its matter-of-fact tone, "and then you'll forget it."

The voice pauses for a moment: a moment in which there is neither sound nor picture. Then an image of a Dristan bottle flashes on the screen for a millisecond, just long enough to be recognized before it disappears. Over the blank screen the narrating voice casually says, "Forget it", ending the ten-second ad.

There is a subtle but slightly superior and knowing tone to the narrating voice, a tone that matches the unqualified assumptions that the voice is making about us – the TV viewers. This is the voice of the advertising industry itself, which has researched and studied us so thoroughly that it knows precisely how we will behave, especially when it comes to the television medium.

This knowing voice first engages our attention by speaking over a blank screen. In the context of television's non-stop imagery, this blank screen is at first a curiosity, especially because the voice is saying "Here's something" when there is clearly nothing. But then, upon this initial paradox, the narrating voice builds a second paradox by completing its statement: "something to forget". We move from curiosity to fascination. We are not used to having a television ad call itself forgettable. What is this something-to-forget that we have still not seen?

The blank screen becomes riveting. We are engaged by its lack. Like this blank screen, the narrating voice is also withholding. "You'll watch it, and then you'll forget it," the voice says, refusing to name this something-to-forget, further delaying the fulfilment of its verbal offering and even slightly humiliating us, with our utterly predictable behaviour that this industry knows so well. Our fascination has become more intimate: triggered by this voice which seems on familiar terms with us. It says "you", perhaps the personal you of intimate address, and it is subtly teasing us: holding back the something-to-forget, holding back the image itself at the same time that it is suggesting it. The lack becomes expectant desire: a desire for the withheld image.

But then there comes a momentary pause: a pause in which there is neither voice nor image. Desire becomes mixed with dread, for suddenly there is nothing: an audiovisual void, a dark and silent screen, an emptiness that momentarily resonates beneath our aroused desire for the image. In this lonely moment of dread and desire, where everything is withheld but the scanning process itself yielding a dark void, we wait in pleasurable agony for the "something to forget", for anything but this nothingness, this pure television void impressing its scanned mosaic on our brain pan.

Finally, thankfully, an image is given: a millisecond flash appears out of the void. It is something familiar, something wonderfully familiar, in the midst of the momentary dread. A glimpse of a Dristan bottle: entirely mundane, ordinarily forgettable, but in this moment of aroused desire and dread, it has filled the nothingness, filled our expectant desire for the withheld image. This flash of the Dristan bottle is pleasurable. "Forget it," says the narrating voice, and on some level, we know we never will.

Pleasure and desire are founded in an urge to repeat, and subsequent viewings of this ad slightly alter but do not diminish the activation of desire in these ten seconds. "Here's something to forget," says the knowing voice, and immediately we recall, though we have been told not to, that first time. Now the momentary pause, where there is neither voice nor image, is filled not so much with desire and dread, as with the pleasurable knowing thrill that the image is coming if we wait for it, wait and ... aahh, it fills the screen, even if only for a moment. And yes, it's the same Dristan bottle, the familiar beloved, forgettable and yet unforgettable.

But there is another key difference between that first time with the Dristan ad and subsequent viewings. The first time, in our state of aroused desire and dread, it was satisfying enough that the image fill the screen only for that brief millisecond of pleasure. But on repeated viewings, we may want the image to linger, to dally, to maybe fade slowly from the screen, rather than disappear again suddenly like a phantom lover. For a more lengthy gaze upon this familiar beloved, we will have to find Dristan in-the-flesh, so to speak. It is there waiting on the supermarket shelf. Our eyes will light upon it and in that flash of recognition, that remembrance of things past, we will reach for one another and nothing will be withheld.

6

Watching for the Magoos

WHENEVER A "NEW" season of TV shows gets underway, you might want to enhance your viewing pleasure by engaging in a pastime that I call "watching for the Magoos". This is an activity which may be either a solitary pursuit or a competitive sport suitable for the entire family. Variations like Mixed Doubles Magoos are also possible. Watching the Magoos is best done while tuned to U.S. shows – but since the Canadian networks and the CRTC itself work hard to bring us all the American programming, it seems somehow wasteful and thankless to not watch at least some of it.

Every new American TV series gets its start as a pilot – a single program that is supposedly a typical episode of the proposed series. The pilot is produced "on spec" by the network's division of program development, or by an independent production company hoping to interest a network in deficit-financing the series. Of the handful of new series that each network offers every autumn (and at less predictable times throughout the year), at least ten times that many competing pilots have been considered by the network's programming department. In other words, a pilot has about a one-in-ten chance of being bought and developed into a series.

Of course, to complicate things, it should be added that several thousand "concepts" compete for the chance of being made into a pilot. And of the one-in-ten pilots that actually get made into new series, 90 per cent fail to survive their first season. (Undoubtedly, the most apt metaphor for this whole process is sexual: like countless sperm fighting their way towards the insouciant egg, most will perish and, just maybe, one will be a hit. But we're getting away from Magoos.)

Given this intense competition and incredible financial investment, as well as the incalculable monetary rewards to be gleaned from scoring in the Nielsen ratings, the TV industry is (as ABC's Grant Tinker once put it) "a nervous business". Obviously, to reduce the twitching and trauma, it's best to find out What The Public Wants. So the first step is to do some concept-testing, long before any money goes into a pilot.

In concept-testing, a carefully selected sample of TV viewers is given a one-sentence plot, as in the following: "A boarding-house run by a female body-builder is mistakenly surrounded by the L.A. SWAT Team. Would you watch that show on TV?" If the sample answers a decisive yes, the concept may advance to the pilot stage. If these potential viewers reject the concept, it may be scrapped or else changed and tested again: "A boarding-house run by a female body-builder for visiting NFL cheerleaders is mistakenly surrounded by the L.A. SWAT team. Would you watch *that* show on TV?" Of course, the intrepid production-researchers may keep tinkering with the concept until they get it right. "A boarding-house run by a secretly gay female body-builder for visiting NFL cheerleaders is mistakenly surrounded by an L.A. SWAT team with herpes. Would you watch that show on TV?"

But once a concept has reached the pilot stage, to the tune of several hundred-thousand dollars invested, that's when the real nervousness begins. To ease the strain, there are the Magoos. Provided by an outfit known as Preview House, the Magoos seem to significantly reduce the valium intake of *Variety*'s executive-producer readership.

□

Preview House, located on Sunset Boulevard in Los Angeles, is a specially-equipped four hundred-seat theatre for testing audience response to TV pilots, finished and rough-cut commercials, feature films, and record albums. Originally called Audience Studios Incorporated when it was founded in 1965 by Columbia Pictures, Preview House is now owned by ASI Market Research Inc. and does a brisk business with more than two hundred regular clients for well over $ 10 million annual revenue.[1] Two of its more notable clients are the ABC and NBC television networks, which are willing to pay about $ 5,000 per pilot to get the Magoos. (CBS has its own form of pilot-testing, which we'll get to later.)

Several nights a week, ASI recruiters telephone a random sampling of Los Angeles residents, or stroll through crowds at public events, super-markets, and tourist spots, and offer a free evening of entertainment. Those who accept the invitation and arrive at Preview House at the appointed hour must first give detailed demographic data about themselves before entering the theatre. Once inside, they are greeted by an Emcee who explains Preview House's "instantaneous response machine". Each viewing seat is equipped with a rheostat dial which the viewer is instructed to turn continually while watching the show. The dial has five different positions marked: very dull, dull, normal, good, very good. Each twist of the dials is monitored by a computer located in a glass booth at the back of the theatre. Technicians there follow a graph, sort of like an electrocardiograph, which traces the overall audience response throughout the screening.

Meanwhile, the computer analyzes eight different demographic aspects of the incoming responses, providing eight different read-outs, moment by moment, of reactions by the audience. Response is broken down into categories such as: men under thirty-five, women, college graduates, fans of Dukes of Hazzard (or whatever program provides the desired comparison), or professional level. By this means a client can find out how a pilot tested with specific sub-groups in the audience.

The first thing every audience each screening night sees is a "Mr. Magoo" cartoon featuring the clinically-blind Mr. Magoo skiing through the Alps. Ostensibly, the cartoon is shown so that the audience gets to practice working the dials before the main programming fare is presented. But since the very same cartoon is shown at every screening, it provides the statistical norm, the common denominator, from audience to audience. A "normal" audience's reactions throughout the cartoon will make the computer read-out needle hit rather predictable peaks and valleys measured on the graph. A less enthusiastic audience for Mr. Magoo will be spotted as registering "a low Magoo", so that subsequent scores for the pilots and commercials shown that evening can be statistically readjusted to provide a more accurate picture. These scores, known as the Magoos, are what the Preview House clients are eagerly awaiting.

After the Magoo cartoon comes the main programming fare for the evening (usually several TV pilots and dozens of commercials). The four hundred audience-members dutifully manipulate their black rheostat dials, sending in the vital flow of data to be instantly analyzed by the computer. A pilot is considered to have done well if its average score is between 5.1 and 6.3 Magoos.[2]

Any scene involving a cute child, sexy woman, car crash, cuddly animal, familiar song or guest star, or all of the above will yield high Magoos. Character development and plot exposition register low Magoos. Dull. Very dull. According to one veteran TV writer, "The ideal dramatic pilot contains nothing but action. If the private eye has to explain something, he'd better be doing something interesting at the time, like walking a tightrope over Niagara Falls or screwing his female client."[3] Says another writer, "If you had a dog in every scene and played an old favorite like 'Raindrops Keep Falling On My Head' throughout the whole thing, you'd get terrific Magoos".[4] ASI vice-president Mark Pinney is convinced of the reliability of the Magoo data. "We're quite accurate in predicting relative Nielsen shares," says Pinney. "Our predictions correlate about 0.9 to the show's eventual share."[5]

While ASI recruiters are out rounding up audiences for Preview House, way across the continent at CBS's Manhattan testing facilities another recruiting team entices unsuspecting tourists into a screening room by flashing cheap ballpoint pens bearing the CBS logo and by talking up an evening of great (free!) shows. The CBS testing system is a bit less complicated than that of Preview House, though no less revered.[6] CBS viewers push a green button when they like what they see, and a red button when they don't. (Got that? Green for like, red for don't like.) The CBS central computer translates these reactions into a series of statistically-averaged red and green blips graphically superimposed onto the pilot videotape at the relevant moments. These red and green blips are then carefully studied by the network executives and program producers, with an eye towards somehow turning all the red blips into green blips during the rewrite.

With the success of Preview House and the CBS Manhattan computer facilities, other research companies have arisen to take the guesswork out of TV production. ERA Research in San Francisco provides galvanic skin response (GSR) data to news organizations who want to test the emotional effects of their anchormen and stories on audiences. After finding the right demographic sample of one hundred people, ERA pays them each $20 to come to its testing hall, where electrodes are attached to the index finger and fourth finger of one hand.[7] As the blow-dried anchorman appears on the screen, wrinkling his brow or smiling as he introduces each plane-crash item, arrival of a dignitary, or local happy-news spot, the electrodes track the changes in the skin's electrical charge for each wired-up viewer, apparently indicating the peaks and valleys of emotional response to each news item or anchor personality. Presumably, the downer-moments indicate where a news organization might consider revamping its agenda, or its personnel.

But TV is a very competitive business, and the U.S. networks are apparently interested in going a bit deeper than the surface of the skin. Imitating the TV / brain research conducted by ad agencies, networks have also tested pilots at the NeuroCommunication Research Laboratory in Danbury, Connecticut. According to a *Newsweek* magazine report, selected viewers at the Danbury lab have had electrodes attached to their scalps to record response to TV pilots by measuring their Beta waves. This activity is not something the networks care to acknowledge.[8]

☐

Watching for the Magoos, then, is quite simple. You need only a scorepad and pencil, plus a working TV set tuned to an American show. Whenever you spot a moment that is obviously a Magoo – something that would make a Preview House audience switch their dials to "good" or "very good", a CBS test-audience push their green buttons, an ERA Research audience sweat with pleasure, or a Danbury audience make waves – you quickly call out "Magoo!" loudly enough so your competitors know the point is yours. The scorekeeper records the point under your name. Best total at the end of the program, or the evening, wins. It's a harmless pursuit that gives the family a semblance of interactive communication. And if you're watching alone, it can keep you from slipping into alpha and setting the sofa on fire.

In the highly nervous world of commercial TV, it doesn't seem to matter that pilots for shows like *Mary Tyler Moore* and *All In The Family* actually scored very low Magoos with test audiences. Better safe than sorry. As TV production costs continue to skyrocket, and as competition for viewers spreads beyond the traditional networks to include pay-TV, cable, satellite-TV, video games, and Lord-knows-what-else, the industry increasingly relies on the Magoos and the little red and green blips and the GSR and the brain-waves to make decisions.

Your pilot got a low Magoo? Sorry. Begin again and concentrate. "A boarding-house run by a secretly gay female body-builder for visiting NFL cheerleaders is mistakenly surrounded by an L.A. SWAT team with herpes and cocker-spaniels. Would you watch *that* show on TV?"

7

The Righteous Stuff:
Evangelism and the TV God

"WHAT AMERICA NEEDS" – the Reverend Rex Humbard hurls the words into the audience like bolts of lightning while little rivulets of sweat form an unholy ring-around-the-collar of his shirt – "is an old-fashioned-Holy-Ghost-God-sent-soul-savin'-devil-hatin' *revival!*"[1]

"Praise the Lord!" the crowd yells back. "Amen! Hallelujah!"

Get set, you sinners. The revival is *on*, and with a vengeance. No more snickering at the prime-time preachers as you nurse your hangover (you godless commie!) on Sunday mornings. No more scoffing at Tammy Bakker's tears, no more irreverent phonecalls to those 1-800 numbers flashing on your TV screen during evangelical broadcasts. This is a real devil-hatin' revival that's underway and, God-willin', there'll soon be a soul-savin', certified TV preacher in the White House.

Why, it wasn't that long ago that Reverend Pat Robertson – you know ol' Marion Gordon ("Pat") Robertson of *The 700 Club,* founder of the Christian Broadcasting Network, leader of the Freedom Council PAC? – well, in November 1985 Reverend Pat Robertson humbly admitted that he was prayin' for a sign: a sign as to whether or not he should make a bid for the U.S. presidency in 1988. And guess what! Praise the Lord! On the seventeenth of February 1986, there it was: Reverend Robertson's face gracing the cover of *Time* magazine! You don't get a more direct sign than that these days. God works in mysterious ways, but an eight-page cover story in *Time* is about the clearest sign anybody would want, especially if they're harbouring thoughts of lust in their heart for the greatest ministry God ever ordained.

And what a down-right respectful cover story it was, too. Those effete, godless, secular humanist reporters must have learned to toe the line. After all, you don't mess around with a conservative Christian fundamentalist revival that has more than twelve political action committees lobbying Congress every day, over $25 million in its coffers, and spokesmen like Howard Phillips (founder of the Conservative Caucus) saying things like: "We must prove our ability to get revenge on people who go against us."[2] Then, too, you don't mess around with a conservative Christian fundamentalist revival that's got its Bible rammed right up the corridors of power, smack-dab into the Oval office.

But you know what? Even Ronald Reagan hasn't been godly enough. No sirree. Not even with his home-spun sermonettes and constant mentions of God Almighty. As Reverend Jerry Falwell says, "In an ideal world, we'd have God for president. Nothing less is appropriate for this nation that God loves above all others."[3]

And that's why Reverend Pat Robertson has been prayin' and talkin' with political advisors and listenin' to those seventy million righteous fundamentalist Christians out there across the God-blessed U.S. of A. – every lovin' one of them with a little nest-egg tucked away that they'd be right proud to put in the collection plate of a man with a direct hot-line to God Almighty Himself, and who's willing to save us from the hellfires and damnation of these godless times.

"We see a virulent humanism and an anti-God rebellion of which blatant homosexuality, radical feminism, the youth revolt and the [UN] Year of the Child, drug abuse, free sex, and widespread abortion are just symptoms," says Reverend Pat.[4] "We have enough votes to run the country. And when people say 'We've had enough', we are going to do it."[5] But you won't find Robertson quotes like that in the *Time* cover story. No way. You put comments like that in *Time* magazine and, quicker than you can kick a hound-dog's ass, you've got your phones ringing off the hook with complaints from evangelists about being quoted out of context, and howls from righteous Christians about the satanic distortions of the media elite, and self-serving rantings from the left-wing commie bloc about something called separation of Church and State.

But as Reverend Jerry Falwell says, "The Founding Fathers did not have in mind separation of God and state, only religion and state."[6] See, sinners? And that's what this old-fashioned-Holy-Ghost-God-sent-soul-savin'-devil-hatin' revival is all about. "In recent months, God has been calling me to do more than just preach," Reverend Falwell said in 1981, "He has called me to take action. I have a divine mandate to go right into the halls of Congress and fight for laws that will save America."[7] But if Reverend Falwell ever lusted in his heart for the U.S. presidency, even while he was proselytizing for Reagan, he blew it totally by telling *Penthouse*:

> If I were going to run for president, that would be bad, that would be wrong. I would be using ... If I were going to run for U.S. Senate in my state, I would be using my influence that I have developed from my ministry to pick up – and I probably could – a majority of voters to go into the U.S. Senate. Okay. That's the [Ayatollah] Khomeini approach. That's wrong.[8]

But Reverend Pat Robertson apparently has none of these hesitations. As he told *Time*, "The only thing for me is, Where would God have me to serve?"[9] Even Reverend Falwell's immediate inferior at Moral Majority, executive director Reverend Robert Billings, religious liaison for the Reagan campaign in 1980, has perhaps better sensed the spirit of the times: "People want leadership. They don't want to think for themselves. They want to be told what to think by those of us here close to the front."[10] And how much closer to the front can you get than sitting at that mahogany desk with the presidential seal?

So Reverend Robertson is saying all the right things: "Deficit spending is neither left wing nor right wing; it is just stupid. Balanced budgets make sense."[11] That has a fine ring to it, the truly righteous ring of a born-again businessman running a $233 million-a-year religious empire. Let the others rant and rave and reveal the Christian fundamentalist agenda: folks like Gary Potter (Catholics for Christian Political Action) – "When the Christian majority takes over this country, there will be no satanic churches, no more free distribution of pornography, no more abortion on demand and no more talk of rights for homosexuals. After the Christian majority takes control, pluralism will be seen as immoral and evil and the state will not permit anybody the right to practice evil."[12] Or folks like Reverend Dan C. Fore (Moral Majority) – "We better forget about all these civil rights which have done nothing but destroy society and create anarchy. They're really not rights at all, but lusts and wants."[13]

That kind of rhetoric might be fine for the pulpits, but in *Time* magazine it would go down about as well as a hairball in your hominy grits. Better to say things like this: "I think that freedom is breaking forth in the world. All the U.S. has to do is to stay strong and to stay the course. That is assuming we don't fall from within with moral decay. If we have a spiritual renewal, which is urgently needed, there is no question that the long-term outlook for the U.S. is very, very bright."[14]

It's real nice wording, Reverend Pat. Sounds like all you're advocating is a wholesome brushing with Aim on the spiritual teeth of society. It even reminds me of the "spirit" ads all over the tee-vee these days: Pepsi spirit, the spirit of Hyatt, Spirit – the American Motors car, the Xerox monks, Datsun Saves ... fits right in, y'know? Who'd ever guess that "staying the course" and "staying strong" would mean things like Operation Blessing: Robertson's $2 million aid to Contra-controlled refugee camps for Nicaraguans in Honduras. Or that "spiritual renewal" is aimed directly at all those virulent humanists, blatant homosexuals, radical feminists, revolting youths, drug abusers, free-sex advocates, the pro-choicers runnin' around out there doin' the devil's work.

□

At the forty-third annual National Religious Broadcasters Convention in 1986, a declaration of support for SDI, or "Star Wars", was released and signed by many TV evangelists, including Jimmy Swaggart, Jim Bakker, and Jerry Falwell. Meanwhile, Reverend Pat Robertson reaffirmed his view that communism will be eliminated from the world by the end of this century. Throughout the 1980s, America's hundreds of religious broadcasters have been expanding the scope of their gospel message to include international politics and foreign policy. This expanded scope is behind the recent name-change of the Moral Majority, to Liberty Federation. "With the name Moral Majority," explained a spokesman, "some people had the impression that the only thing we could be concerned with was pornography or abortion. But the name Liberty Federation provides us the opportunity to address issues like Soviet expansion without people balking."[15]

"We religious broadcasters," says Reverend Robertson, "are a symbol that a profound spiritual renewal is taking place in our country."[16] Well, you couldn't ask for a nicer symbol: a $2 billion-a-year industry. Come to think of it, you couldn't ask for a better symbol for God-blessed-America itself! Why, everybody knows that good old American TV is just the best goll-darned television in the whole world. And everybody knows (or they ought to) that America is the Promised Land chosen by God Almighty Himself! As Republican congressman Jack Kemp of New York explains, "God was the author of the Declaration of Independence".[17] So your prime-time TV preacher, reaching across the airwaves to gather the nation and whole world into one big, beautiful, American TV congregation, has got to be the epitome of the righteous, God-given Manifest Destiny of the U.S.A. As Reverend Jerry Falwell told a Toronto audience in 1982, "We are all 'fellow Americans'" in the eyes of the TV / God.[18]

You see, sinners, what you gotta understand is that this old-fashioned-Holy-Ghost-God-sent-soul-savin'-devil-hatin' Christian fundamentalist conservative revival just couldn't happen without your television set. "With electronic evangelism," says Reverend Rex Humbard, "we go beyond the walls of individual churches and we encompass all faiths. It is the instrument by which you reach most of the people in the world – half of whom can neither read nor write."[19] Reverend David Mainse, of Canada's own *100 Huntley Street*, makes it more clear: "When God created the universe, He took into it the laws of communication that are linking peoples today. The earth is prepared to receive messages. There are televisions and radios in homes. The Holy Spirit is acting now through the medium of television simply because the time has come, the world has been prepared."[20]

You can see it in the TV cathedrals that rise up into the heavenly airwaves, glorifying the Almighty TV / God Who has His finger on the cosmic channel selector. Reverend Humbard started the trend with his Cathedral of Tomorrow – the first church designed as a TV studio, built in 1958. But Reverend Robert H. Schuller and his *Hour Of Power* were not to be outdone. By 1980 Reverend Schuller had built his Crystal Cathedral, with ninety-foot doors that open to reveal twelve massive fountains (one for each apostle, don't you know) and an eleven-by-fifteen-foot Jumbotron video screen. Over at the recently troubled PTL Club, the once-Reverend Jim Bakker similarly built his TV station to be a huge replica of a colonial church. Its towering steeple houses a one-hundred-foot broadcasting antenna and its nave is filled with elaborate broadcasting facilities to reach 180 TV stations across the United States and Canada, as well as fifty-three other countries around the globe. Before his sadly adulterous departure from his TV ministry, Bakker had long before admitted his true lust: "I love TV. I eat it. I sleep it," he once wrote. "My specific calling from God is to be a television talk-show host."[21]

Reverend Pat Robertson's approach has always been more born-again business-like: putting the money ($233 million a year) from his Christian Broadcasting Network into renting a twenty-four-hour-a-day satellite transponder on Satcom IIIR, purchasing four UHF TV stations and five FM radio stations, and buying time in 185 TV stations across the continent. "We've got some of the best TV equipment in the world," says Eldon Wyant, manager of CBN's network relations, "and people from ABC will be down to study our new Italian lighting system soon."[22]

With the Jesus network comprising over two hundred full-time religious TV stations, 1,134 radio stations, and the expanding cable and satellite hookups, the TV Holy Spirit is broadcasting to over sixty million North Americans on a daily basis, and to countless millions across the planet. Praise the Lord! As sociologist Jeffrey Hadden, author of *Prime Time Preachers,* says, the TV evangelists "have greater unrestricted access to media than any other interest group". The Christian right "is destined to become the major social movement in America."[23] Author Jeremy Rifkin believes that has already happened: "A close look at the evangelical communications network ... should convince even the skeptic that it is now the single most important cultural force in American life."[24]

To some extent, the expansion of cable-TV across the United States during the 1970s accounts for the expanded ministries of the televangelists, many of whom had previously been known only on Bible-belt stations serving small regions. Not surprisingly, behind that now broad-based TV ministry there is a particular and enthusiastic relationship to electronic technology that makes right-wing televangelism merely the leading edge of the New Right and, in a larger sense, simply the more virulent expression of the dominant, patriarchal culture. That ministry could not exist without the computer (crucial for its direct-mail solicitations), satellite technology (necessary for a global reach), and (obviously) television. All of these technologies are necessary to construct the pseudo-communities by which the movement is financed.

Marshall McLuhan recognized that television is like the traditional concept of God, whose "center is everywhere and whose margins are nowhere".[25] All-seeing, omniscient, all-pervasive, all-encompassing, television comforts us in our loneliness and is always there in our time of need: ministering and giving unceasingly for "free". As McLuhan protégé and media guru / adman Tony Schwartz, author of *Media: The Second God*, has stated:

> Ask anyone raised in the religious traditions of the Western world to describe God and this, with idiosyncratic variations, might be his answer: "God is all-knowing, and all-powerful. He is a spirit, not a body, and He exists both outside us and within us. God is always with us because He is everywhere. We can never fully understand Him because He works in mysterious ways." In broad terms, this describes the God of our fathers, but it also describes electronic media, the second god, which man has created.[26]

How potent, then, to combine this god-like medium of television with outright evangelical content: especially the emotional words and the direct gaze of the prime-time preacher. As one convert to Reverend David Mainse's *100 Huntley Street* puts it: "I was feeling really desperate, and suddenly I saw David look right out of the screen at me and say, 'I know there's someone out there who is aching, who has been delaying the day when she lets Jesus into her heart, who has been putting off the Lord. But today is the day.' And I just felt that the Spirit was talking to me in my heart, that it was me that David meant."[27] Obviously, in this sense, televangelism is not different from the television industry itself, as writers

Margaret O'Brien Steinfels and Peter Steinfels observe: "If you accept Jesus, you will enjoy immediate relief from suffering. Success, prosperity, and earthly happiness will be yours. This presents an odd contrast to Jesus' message, but it bears more than a faint resemblance to the run of TV commercials."[28] In line with the TV / brain phenomenon, the feelings triggered in the right hemisphere can be labelled with Jesus or Anacin. In either case, however, the resonance of the mass medium is necessary.

□

To a significant extent, we must consider televangelism an expression of a larger ideology that could be termed "reactionary modernism". This ideology fully accepts the technological advances of modernity at the same time as it looks back with a yearning nostalgia to an earlier era. Reverend Falwell states:

> It is easy for people today who are violating God's law and man's law to ridicule those who oppose them by simply saying, "That fellow's repressive; he is suggesting a return to where America was fifty years ago, morally." That is exactly what I am proposing, morally. Not technologically. I certainly am very much a progressive person. If you come to our ministry and visit our complex and see the kind of innovative and up-to-date procedures we use in everything we do, you'd never accuse me of being against progress. But there's a vast difference between technology and theology.[29]

Unless your theology is technology on some level. I would argue that this is the case not only with the televangelists, but throughout patriarchal culture, with the projection of the God-archetype onto technological advance as "salvation". We might compare Falwell's statement with another made fifty years ago, in that era to which reactionary modernism would like to return morally. Jeffrey Herf, in his book *Reactionary Modernism: Technology, Culture, and Politics in Weimar and the Third Reich*, quotes from a speech by Joseph Goebbels in March, 1939. Goebbels's words opened the Berlin Auto Show:

> National Socialism never rejected or struggled against technology. Rather, one of its main tasks was to consciously affirm it, *to fill it inwardly with soul*.... While bourgeois reaction was alien to and filled with incomprehension, if not outright hostility to technology, and while modern skeptics believed the deepest roots of the collapse of European culture lay in it, *National Socialism understood how to take the soulless framework of technology and fill it with the rhythm and hot impulses of our time.*[30]

In our postwar culture, devoid of viable communal containment for spiritual impulse and expression, we are encouraged to project that spirit into technological systems. Perhaps not surprisingly, while right-wing televangelists fully support "Star Wars", they are seemingly prepared for a nuclear Armageddon. As Reverend Falwell says: "From Genesis to Revelations the message is clear. There will be a nuclear holy war over Jerusalem and the Russians will come out second-best ... if we are ready for it. The issue is survival. Jesus was not a pacifist."[31] Alan Crawford has characterized the ideology of the New Right as "the Politics of Resentment".[32] That resentment is growing so strong, and has so many targeted enemies that it wants to wipe out, that the purge of a nuclear holocaust holds considerable attraction: especially when it is defined as "winnable".

When the TV Holy Spirit starts telling those seventy million fundamentalist Christians that it's time to put a TV preacher in the White House – someone who can make sure that everything God disapproves of (ERA, rock music, sex, abortion, welfare, détente, defence cuts, homosexuality, liberalism, pluralism, secular humanism, radical feminism) will be stamped out with the righteous indignation of an Old Testament prophet – then it will take more than a flick of the television dial to ignore this revival. As Reverend Robertson says, "We have enough votes to run the country. And when people say 'We've had enough', we are going to do it."

In the meantime, all presidential candidates must present themselves as paragons of fundamentalist morality. Indeed, sexual morality has become *the* "issue" of political leadership. As David Frost, the British journalist at work on a series of TV interviews with 1988 presidential candidates, admits: "Today, the question on everyone's mind is, 'Have you ever committed adultery?'"[33] Praise the Lord and pass the ammunition.

8

TV News: A Structure of Reassurance

"News" are actually "olds," because they correspond to what one
expects to happen.

JOHAN GALTUNG and MARI HOLMBOE RUGE

WATCHING CBC's network news show, *The National,* I'm often captivated
by the sheer technological brilliance of the production. *The National,* like
all other network news shows, is a complex interweaving of disparate elements – filmed reportage, live studio coverage, rear-screen graphics, minicam transmission, satellite feeds – all combined into an apparently seamless whole. Add to this complex collage the sophisticated computer animation that *The National* uses to open and close the show, and the result is a
sense of television technology taken to its limits.

No other TV genre brings together such a range of technological competence. Arguably, the network news show is essentially a showcase for the
latest in electronics hardware and a celebration of television itself. Seen in
this light, the recurring structure of the nightly newscast reveals an interesting ideology at work behind the overt content.

Over the past thirty-five years, the technological goal of television news
has always been to achieve more up-to-the-minute coverage of events on
location. Each advancement in the television apparatus can be understood
in terms of this goal, especially the development of ever more portable,
light-weight cameras. By the early 1970s, the introduction of ENG (elec-

tronic news-gathering) technology seemed to herald the approach of the ideal. The small minicam cameras are easily portable and produce a sharp image. Better yet, ENG equipment, unlike film cameras, simultaneously feeds electronic pulse back to the studio for immediate transmission or for storage on tape. With film, there is always a delay of several hours before the material can be broadcast. By the mid-1970s all North American network news agencies had invested in ENG technology, not only for competitive reasons and the implicit status accompanying every technological purchase, but also because the equipment was the latest breakthrough in achieving the overall technological goal of TV news itself.

Behind this desire for more up-to-the-minute coverage on location, there is, perhaps, a deeper motivation. As Wallace Westfield, former executive producer for NBC, said:

> Television people have always been worried and fearful of a comparison with print people. It started really in the fifties when television news became a fact. It was in a fifteen-minute form on a daily basis and, I think, in those days the broadcast journalists were always somewhat embarrassed. They felt that they suffered by comparison with print. I think this sort of set the mode for broadcast journalism.[1]

It has been common knowledge for decades that TV news does not achieve the depth of analysis possible in print. The verbal portion of a network news show would fill less than half a page of a newspaper. Given this unfavourable comparison, TV news has always sought its own uniqueness. In almost defensive fashion, each technological advancement has been an attempt to stake out television's specific news terrain. Simply put, the mode set for broadcast journalism was an intense fascination with the technology of the medium.

We can sense this vividly in a transcript from a *See It Now* program broadcast on November 18, 1951, and hosted by Edward R. Murrow. The occasion was the first TV linkup, through cables and relay stations, of East and West-coast U.S.A. Murrow states:

> We are, as newcomers to this medium, rather impressed by the whole thing; impressed, for example, that I can turn to Don Hewitt and say: Don, will you push a button and bring me in the Atlantic coast? Okay, now San Francisco, could you use what you call, I think, a Zoomar lens and close in on the bridge a little? We, for our part, are considerably impressed. For the first time man has been able to sit at home and look at two oceans at the same time. We're impressed with the importance of this medium. We shall hope to learn to use it and not to abuse it.[2]

In our present era of satellite telecommunications, the excitement expressed here may seem oddly quaint. Four times Murrow says he's "impressed", revealing a bedazzlement that cannot be masked by the sudden solemnity of his closing lines. But what is of interest here, for our purposes, is the specific object of Murrow's fascination: the simultaneous live transmission of on-location visuals. Had the images of the two oceans been filmed images, made earlier in the day on each coast and then linked up for simultaneous transmission through the cables and relays, the reporter would certainly have been less impressed. In other words, it was not the linkup of the coasts which so bedazzled Murrow, but the linkup of live on-location transmission in real-time. He proves this by asking that the "Zoomar lens ... close in on the bridge a little". It is this simple request, followed by an answering change of frame, that established, once and for all, the unique terrain of the medium. No wonder a dedicated TV newsman like Murrow was so bedazzled. Television had found its "news beat" – live on-location transmission in real-time: a more impressive terrain technologically than that of either print or film.

Yet the irony of TV news is that its dictates *as a program* overshadow its prestigious capabilities. In actual fact, the only part of a TV news program that is transmitted live in real-time is the image of the anchorperson in the studio. Even with ENG technology, we seldom see an on-location item broadcast live. The contradiction – between technological potential and programming demands – results in several repercussions for the news.

Because the news is a program like any other, it must fit within the broadcast schedule in its allotted time period. Therefore, each item on its agenda must be timed and slotted into the overall rhythm of the show. Reality, of course, is not so neat. As the only part of the show transmitted live in real-time, the studio anchorperson, therefore, has certain crucial functions. Primarily, he or she is the signifier of live coverage.

A nightly news show is a complex blending of myriad time-space frames. Of the twenty or so individual news items on the agenda, there may be a filmed item shot six hours earlier in the Middle East, another filmed in Europe, an item using hour-old ENG coverage from a downtown city area, or a satellite feed from another network earlier in the evening. As the signifier of live coverage in real-time, the anchorperson must confer the aura of "presentness" on everything else in the show. He or she must introduce each news item and thereby (as the word implies) "anchor" it within the space-time frame that the anchorperson represents. Only then can the screen be relinquished to a previously filmed or taped segment. The image of the anchorperson brackets every item, conferring upon it the resonance of live transmission in real-time that he / she signifies.

Although reporters are not allowed to usurp the special status given to the studio anchor, their news items must approximate it by providing "same-day pictures" of newsmakers. These "same-day pictures", whose content is no more meaningful than a still photograph would have been, are necessary to remind us of the special promise inherent in television's unique news terrain. Though the promise is fulfilled only by the anchorperson, the "same-day pictures" reinforce a special sense of television as a news organization, a sense that TV gives us what Stuart Hall calls the "having-been-there" of news. Todd Gitlin writes:

> At CBS News in November 1976, I saw a report come in by satellite of Leonid Brezhnev's arrival in Belgrade for a meeting with President Tito. The correspondent had flown in from Paris and framed the airport scene like every other diplomatic and presidential airport arrival scene ever broadcast. Nothing was revealed. The retired correspondent Alexander Kendrick maintains that this conventionalization of foreign coverage is a consequence of satellite technology: the network can get a same-day report now, and so loses interest in deeper (and more expensive) on-the-spot coverage of all but the most riveting continuing foreign stories.... I would say that such technology meets the organization's need to process packaged information quickly and cost-effectively; the technology is developed, promoted, and made to appear inevitable toward that end.[3]

One result, as Michael Arlen, TV columnist for *The New Yorker*, told me, is that: "There is an enormous variety of events being presented all in a kind of illusion of presentness, as if they all took place this evening. Now and then a television news organization will make an enormous and special effort to connect an item back to something, but it's always a very special effort. The rest of things are just simply floating in the present."[4]

This illusion of presentness, built into the structure of the news program through the bracketing function of the anchorperson, works to convey an ideology in which the present frames and brackets the past. Individual news items are treated as discrete and separate entities, with little or no relation to other items or to a larger historical context. The illusion of presentness conveys the sense that events take place in a vacuum and are entirely self-contained. An ideology in which the present is seen as presiding over the past is somewhat of a reversal of reality, wherein the past gives birth to the present and explains it. But as an ideology this illusion of presentness is useful to TV's purposes.

Without historical context, information becomes bits of trivia. Viewers may find these bits "interesting" but may be unable to connect them to each other or to anything else. Without context, viewers may accumulate information and data, but have no real understanding or insight into why something is happening or what is behind an event. Moreover, without historical context, individual news items will simply be given our own personal context: that is, we will anchor the data within the limited confines of our own knowledge, memories, even our own fears and prejudices.

Another more revealing possibility is suggested by the content inherent in TV's desire for "same-day pictures". The events most amenable to this desire are those that can be planned for in advance: the arrivals and departures of statesmen, press conferences, meetings of heads of state, photo opportunities arranged for the press – the so-called "pseudo-events" that characterize so much of journalism these days. As Michael Arlen puts it:

> Basically, I think that network news is almost entirely a news of important people talking to other important people, or about important people. It's a news of institutional events. It is bureaucratic. By and large, network news goes out of its way to present a pageantry of officials everywhere making official statements about official things.[5]

In place of wider historical context, TV news substitutes an illusion of presentness populated by officialdom. In other words, history is replaced by institutions as context. As viewers, in our efforts to understand why something is happening, we may rest assured that, although a particular event might seem inexplicable to us, presumably somebody else knows the necessary background and context for the information: undoubtedly one of the many officials we see arriving and departing, shaking hands, and making official statements.

Thus, TV network news continually reassures us of the viability of society's official institutions. Since television itself is the most eminent and omnipresent of official institutions, it has quite a stake in this reassurance function.

□

Studio news anchorpersons are the officials *par excellence*. In the structure of the news program, their role is a mirror image of officialdom in the wider society. That is, their statements carry more authority than anyone else's, given the structure of the program. As the signifier of live coverage, their presence is crucial to the show, whereas individual reporters (and news stories) may come and go. Almost a full year in advance, viewers were being prepared for the retirement of Walter Cronkite as CBS anchorman. Over the ensuing months, we could, in effect, watch his immediate successor, Dan Rather, take on the anchorman "aura". Presumably, through such advance notice, no undue rupture would occur in our perception of the signifier of live coverage.

Moreover, only the anchorperson is invested with the special status that television technology claims for itself: the importance and immediacy of "now". As Arthur Asa Berger, author of *The TV-Guided American,* says of Cronkite, "He merges with the news, his presence has come to be regarded, by many people, as an indicator of the significance of any event."[6] Before his retirement, Cronkite's presence on a TV special often meant not only that the coverage was important, but also that the transmission was occurring live in real-time. Carrying this aura with him from the news studio set, he could confer it upon other TV programming. On CBC, anchorman Knowlton Nash has come to signify this same combination of important coverage and live transmission. He, too, appears on special event programming, conveying a special status (newsworthy and live) to the show.

Thus, the network news anchorperson signifies the institution of television itself. Former president of NBC News Reuven Frank has called the anchorperson "a fixed center in all the chaos ... a reference point [for the audience], something dependable, like the familiar typeprint of your favorite newspaper."[7] As the only medium that can bring us live on-location transmission in real-time, television as an institution seems larger than any and all other institutions. It can show and comment on them all, overseeing and framing them within the boundaries of its screen, bracketing them all within the illusion of presentness that the technology claims as its own. Indeed, everything else is past, given television's usurping of the now.

It has become commonplace for news items to incluae images of television crews at work covering events. On our screens we see a cluster of camera, lighting, and sound personnel busily pursuing the ostensible subject of the news story. The old 11:00 p.m. format of CBC's *The National* used to ritualistically end with the image of a studio camera crew at work in front of the news desk. This self-reflexive style does more than suggest the newsworthiness or importance of a particular figure or event. In a larger sense, this stylistic convention proclaims the institution of television at work and inscribes its own image within its process. "The television of the late 1970s and early 1980s concerns itself with *TV qua TV,*" writes Edwin Diamond.[8] This self-reflexive style refers us to the higher-level system of television itself as the predominant institution.

President of CBS News Richard Salant once commented: "Our reporters do not cover stories from *their* point of view; they are presenting them from *nobody's* point of view". The "nobody" referred to here must surely be the institution of television: that disembodied, all-encompassing entity that embraces everything in its "view", showing us not only the world, but itself showing us the world.

As viewers adrift in a sea of information, swept away by a deluge of information without historical context, we are meant to find reassurance in the fact that there is one institution that sees and frames all others. Whatever ripple of disquiet, whatever wave of potential disruption may sweep over the status quo of other institutions, we know, by the very fact that television is showing it all to us, that all is well: embraced in the all-seeing gaze of the television eye and the calm visage of the studio anchor.

9

An Electronic Marriage

IT'S TERRIBLY EASY to forget that the role of television is neither to inform nor entertain us, but to sell "heads" to sponsors. Indeed, the medium depends on this forgetting, this tendency to perceive TV as merely a servant to our pleasure. But, in fact, television is demographic, not democratic. As Les Brown writes in *Harper's* magazine:

> How much an advertiser pays for a 30-second spot depends not so much on how many viewers tune in but on the *quality* of those viewers, affluent young adults being preferred. Women between eighteen and forty-nine, for example, watching during prime time on Fridays, sell for $16.50 per thousand; older women and teenagers, who buy less, sell for less.[1]

So-called "public television", such as PBS and CBC, is similarly dependent upon such head-counts to assure sponsors that the audiences delivered to them are "quality" viewers, that is, viewers with purchasing power.

Fittingly, most histories of television assent to the fact that the first televised image in North America was a dollar bill, placed in front of a TV camera by one of the medium's inventors. With that historic and resonant image in mind, we can turn to a brief consideration of what is perhaps the ultimate power behind television throughout the world: The A.C. Nielsen Company.

As the leading market researcher and broadcast-ratings service, the

A.C. Nielsen Company has a profound impact on those two key trivial pursuits of the postmodern world: shopping and watching TV. Through its careful and unrelenting accumulation of statistical data on both product and program performance, in stores and on the home screens, the A.C. Nielsen Company has become the ultimate arbitrator for consumption world-wide. As of 1980, its services covered only twenty-three countries.[2] But its impact extends somewhat further.

Arthur C. Nielsen started his statistical consulting firm in 1923, selling "performance surveys" to manufacturers who wanted to know what consumers thought of their products. In the late 1930s, Nielsen added a radio ratings service to his business, depending on listeners to keep a diary of what they listened to each week. But Nielsen soon discovered that his diary-keepers were not terribly efficient or reliable. Besides neglecting to be diligent about keeping track of their tuning, Nielsen's team of diarists, more often than not, skewed their listing of shows to those with snob-appeal – as though they wanted to impress him with their good taste, if only on paper.

Thus, Arthur C. Nielsen must have been elated when two professors at the Massachusetts Institute of Technology invented the "audimeter" in 1936. This was a machine that could be attached to a radio and automatically record the station and time tuned-in. Nielsen bought the rights to this invention and converted it for television in the 1950s. Through the magic of this new technology, the chosen "Nielsen families" couldn't report that they'd been watching Edward R. Murrow when in fact they'd been tuned to *Ding-Dong Schoolhouse*. The audimeter cooly eliminated such discrepancies.

Since the 1950s, an audimeter (about the size of a cigar-box) has been placed annually in 1,200 U.S. households, carefully selected according to specific demographic requirements. For example, until 1976, the chosen Nielsen families included no poor households and only a light sampling of Blacks and Hispanics.[3] These audimetered families (their identities always a secret) have provided the viewing data on which the televison ratings are based. Each audimeter continually monitors the tuning activity in each household: the specific TV channel and exact time of viewing throughout the day. During the night, this information is fed to the central computer in Dunedin, Florida, which correlates all the viewing data, providing household demographic breakdowns per program and determining the exact rating and share of available viewers for every network program. Accordingly, sponsors flock, or not, to any given show, and the network executives engage in great gnashings of teeth while they wait for the "over-nights".

Thus, NBC's *Miami Vice* – attracting the desirable high-spending male

Yuppie audience – could, as of autumn 1986, charge $165,000 for each of its thirty-second advertising spots, higher than the average rate of $118,840 for thirty seconds of prime time. Those spots would be filled by car and high-tech advertisers rather than by detergent or food advertisers, who preferred spots on a show like CBS's *Dallas* ($195,000 for a thirty-second spot), watched primarily by female viewers.[4]

For years the Nielsen Company has also monitored, or "tracked", specific items in supermarkets, department stores, liquor stores, and drug stores: comparing performances of brand against brand, special displays as compared to ordinary displays, the effects on sales of the addition of a new ingredient (like "lemon") to an old product, or the results of a new package or special promotion for a product. Such detailed product information is stored in another central computer (this one in Northbrook, Illinois), and its correlated data has become the holy entrails studied by marketing shamans.

During the late 1970s the A.C. Nielsen Company began simplifying its work by issuing special I.D. cards to another sample of U.S. households that shop at specific stores electronically equipped to keep track of their purchases. This information, less anonymous than item-counts on store shelves and more demographically revealing, is also computer-correlated and sold to clients like Procter and Gamble for eager perusal. It is obviously important for a marketing agent to know whether college-educated women between the ages of eighteen and forty-nine, with an upscale income, prefer baby-powder-scented super mini-slim pantyliners over unscented regular deodorant maxi-pads or contour multilayered extra-dry regular-size maxi-liners. Upon such data does the economy rise and fall.

Until the summer of 1985, Canada was shamefully un-audimetered: its television ratings and product performances were determined by the primitive methods of surveys and interviews and those unreliable diaries that Arthur C. Nielsen came to loathe. But that summer, the Nielsen people decided to boldly sweep the nation into the last quarter of the twentieth century by introducing electronic TV-metering. Even more daring, the A.C. Nielsen Company of Canada (located in Markham, Ontario) recognized that this technological move might be the occasion for great continent-wide statistical gains, especially if it could use the same sample base for deriving *both* viewing and purchasing data. Accordingly, Toronto became the North American testing-grounds for a revolutionary new concept in market research called "single source data collection".

This concept brings together the electronic measurement of what people are watching on TV with what they are buying in supermarkets (also electronically measured) as a result. As Nielsen vice-president Everett Holmes says, "The joining together of these two measurements will enable marketers to measure the entire sequence of consumer behaviour – from what people view to what they do."[5]

Thus, up to two thousand Toronto households have been equipped with both the Nielsen "people meter" (an updated version of the audimeter) and the special electronic I.D. cards to use in specified supermarkets. These new Canadian Nielsen families are providing a wonderful new data base through which advertisers can determine the effectiveness, and the effective placement in TV programs, of their commercials.

The "people meter" is clearly state-of-the-art technology. It is a small microprocessor attached to the home TV set and the telephone line, with a hand-held remote keypad that contains eight individually assigned buttons – enough for each member of the Nielsen household and possible guests. This system accurately monitors all TV-viewing by each individual: an important modification to the old audimeter, which only reported overall household viewing. (Thus, little Johnny must remember to push his assigned button on the keypad or the computer will be unhappy.)

Every three seconds, the people meter scans the tuning activity of the set and detects precisely which station (up to two hundred capacity) is being watched and by whom, and it recognizes whether or not that station is being received off-air, by cable, or via satellite dish. It also detects "zapping" – the skipping of commercials through the use of the remote channel selector – and informs the advertiser which portion of the ad (if any) held the attention of which viewers. Of course, the same thing is true for the programs: the people meter, scanning the TV set every three seconds, lets a program-maker know where a significant portion of viewers (and which ones demographically) zapped to another channel. The system also monitors individual use of VCRs, video games, and personal computers hooked into the TV set. Every twenty-four hours, during the wee hours of the night, all this information in each people meter is automatically transferred through the phone lines to the central computer.

These same Toronto Nielsen families shop at designated supermarkets equipped with electronic checkout scanners that track the Universal Product Codes on product labels – those ubiquitous black lines that suddenly appeared on everything during the 1970s. The checkout scanner matches the list of purchases with the electronic I.D. card (like a credit card) provided each household, and sends this purchasing data to the central computer to be correlated with that family's TV-viewing data.

According to Wendy Miles, Account Manager for A.C. Nielsen's Media Research Group, Toronto is "ideal" as the testing-grounds for single source data collection because "Toronto is the largest market in the country, and it has one of the most complicated TV markets in the world".[6]

With more television channels available in Toronto, and from more sources (off-air, cable and satellite) than virtually anywhere else in the world, the city is a market-researcher's dream come true. Since 1985, the company has also begun people metering homes in Denver, Colo., and Springfield, Mo.

In case you hadn't guessed, A.C. Nielsen's single source data collection is the most technologically sophisticated system for monitoring consumer behaviour that has ever been attempted. It is also, to my way of thinking, the most significant electronic marriage between television and the computer that has yet been consummated. We would probably do well to place this event in the context of other systems of computer monitoring of human beings, since there is every reason to believe that eventually such data bases will interface, if they are not doing so already. The gains from such elaborate correlations of personal data are evident to all marketers, and not surprisingly, the on-going results of the Toronto testing-grounds have been closely followed by the entire field of political and product polling and marketing across the continent.

□

In this sense, it would be wise for us to consider the words of one Richard Wirthlin: "There is a tendency in our increasingly complex and highly technological society to forget that American democracy is less a form of government than a romantic preference for a particular value structure."[7]

While most of us would be shocked at the cynicism evident in this statement, we might be even more surprised by the identity of its speaker. Richard Wirthlin is the personal pollster and political strategist of Ronald Reagan. The statement appears in the 1980 "campaign Bible" that Wirthlin co-authored to engineer Reagan's presidential election. In other words, the passage was intended to be read only by Reagan himself and by the team of experts who would execute the campaign strategies Wirthlin had designed. As such, it reveals the operations and philosophical base-line of the new set of powers that have converged over the past twenty years: an elite class of technocrats whose expertise combines technological efficiency, television know-how, and in-depth marketing research techniques with a no-nonsense view of democracy as little more than a set of "romantic preferences".

Wirthlin joined Reagan's team of experts during the 1970 California gubernatorial race in which Reagan was re-elected. At that time, Wirthlin began perfecting his "Political Information System" (PINS) – a mass of complex demographic data analyzed and correlated by computer to yield specific target groups of voters. As Will Ellsworth-Jones and Roland Perry write:

> For twenty years Wirthlin has computer-filed his own polling data in the hundreds of campaigns he has run for Republicans, along with quantities of census figures, information from 37 federal departments, voting history figures from every county, and extensive market survey work for scores of American businesses.... Wirthlin's computers can provide him in an instant with the political preferences and behaviour of 110 categories of the American electorate.[8]

In 1970, this computer targeting was a pioneering strategy, used by Reagan's team to shape specific political ads to match the interests and concerns of individual audiences. The strategy has since become commonplace, perfected and refined into "psychographics": in-depth research into the lifestyles, hopes and fears, fantasies, personality traits, emotional needs, prejudices, ideas for candidates' behaviour and appearance, beliefs and dreams of target groups as specific as individual city neighbourhoods.

This data is so massive and detailed that only a computer could sort through its intricacies: matching ideological mind-sets with income-levels, lifestyles, political preferences, media choices, hopes and fears. Such correlated data allows other members of the team of technocrats to tailor speeches and ads quite specifically to key target audiences and to place those ads in the appropriate TV time-slots and publications – no doubt after having studied the A.C. Nielsen demographic data regarding viewer-composition per program. This information further allows politicians (or their speech-writers) to incorporate the appropriate buzzwords, clichés, catch-phrases, and allusions that will speak to an audience's "romantic preferences" and notions about "democracy".

Reagan – with his soothing grandfatherly voice and demeanour, his jokes and "aw shucks" delivery, his allusions to movies and to some nostalgic, vague American past – has relied on precisely this widespread "romantic preference for a particular value structure" that has come to replace democracy as a form of government. Ellsworthy-Jones and Perry, writing in the spring of 1984, noticed that two of Reagan's January speeches included the word "God" no less than thirty-four times; and in one memorable TV address, Reagan used the word "peace" a total of forty-seven times.[9] Wirthlin, tracking the percentage of citizens who perceived Reagan as "dangerous and uncaring", noted that following such speeches this grouping had dropped from 50 per cent to 44 per cent. In an era when the manufacturing of audience perception is paramount, the designed TV image counts more than actual decisions Reagan has made.

Along with targeting to find out voter-consumer preferences, tracking is another computer technique borrowed from market research for political use. Wirthlin also pioneered the use of this technique in the 1970 Reagan gubernatorial campaign. Tracking means conducting frequent polls before and after specific events, speeches, and broadcasts to determine the effect on citizen response. Tracking has become so widespread that it is now used continually, while a team's client is in office, to monitor his or her performance. "It's like turning on the television set," says Wirthlin. "We leave it on all the time. We don't take our finger off the pulse."[10] After a poll reveals citizens' perceptions, a politician can take steps to alter those perceptions through a timely speech, an appearance on prime time, a press conference, a photo-opportunity including a popular celebrity, or some other activity. Just as A.C. Nielsen's clients need to know from the company's computer-correlations whether or not a change in packaging, display, or the addition of "lemon" has altered sales, so the political team is equally concerned with tracking on-going performance.

Not surprisingly, President Reagan has held the fewest press conferences of any president in half a century. As Lynne Olson writes: "His people want Reagan to be the Great Communicator, but they want him to communicate directly to the people, primarily through television appearances, rather than through the fault-finding filter of the press."[11] From the Christmas season of 1985 through the Shuttle-disaster of January 1986, Reagan appeared (unfiltered) on prime time regularly, including such programs as "A Tribute to Dutch Reagan" and "Christmas In Washington" – guest appearances that seemingly endeared him to the viewing public and removed him from the realm of politics.

By 1980, Wirthlin had added another computer-technique to his political repertoire: computer-simulation testing, which allows the technocrats to run a variety of scenarios for action, or possible ideological positions to be advocated, in order to see how the public will respond. This technique was useful in the 1980 campaign for predicting citizens' response to various possible actions taken by Carter. The Reagan team could then devise in advance their own strategies for whatever scenario might arise.

As Michael Posner reported in 1981, Reagan's "White House advisors possess the most finely tuned public relations antennae ever to extend from the Oval Office, and his presidency is fast-becoming the most stage-managed in history."[12] This should remind us that, although much is made of Reagan's former career as a movie actor, little is said about the fact that he was a public-relations man for both General Electric and Borax in the early days of TV. This bit of trivia took on interesting overtones as soon as he assumed office as president. He immediately surrounded himself with a team of public-relations pros: his Deputy Chief of Staff, Mike Deaver, was a partner with the Hannaford Company – a P.R. firm hired to improve the image of the right-wing Guatemalan government; his Deputy Assistant for

Policy Development, Ed Gray, was director of public relations for the San Diego Federal Savings and Loan Association; his Director of the Office of Cabinet Administration, Craig L. Fuller, was a former vice-president of the Hannaford Company; his first press secretary, Jim Brady, held key P.R. posts with federal agencies and with the Republican Party; his Deputy Press Secretary, Larry Speaks, was a vice-president of Hill and Knowlton, the largest independent public-relations firm in the United States.

Given this much P.R. expertise in the Oval Office, and in politics at large, we cannot ignore the extent to which our perceptions are engineered and finely tuned according to specific political needs. Such expert engineering would be impossible without the twin technologies of television and the computer: the computer because of its abilities to correlate massive amounts of data; TV because of its unchallenged and hegemonic position in daily life. As the linkups between these two technologies proceed, we are confronted with an ever more refined system for monitoring our behaviour – "from what people view to what they do", as the A.C. Nielsen vice-president puts it.

The next stage will no doubt be the widespread use of what's called "two-way, interactive cable TV", whereby a polling computer is connected to home TV sets equipped with a remote keypad of response buttons to answer yes / no and multiple-choice questions. Such a system has already been tested in both Canada and the United States and is being lauded as a "democratic" new means for getting instantaneous feedback from citizens: their opinions, preferences, and concerns. Politicians could also "interact" with viewers in so-called "TV townhall meetings", and voting itself could be conducted through this system.

The implications of such a technological linkup are awesome, given the extent to which we already rely on television for information and the extent to which the medium shapes perceptions. Moreover, the defining of complex issues through the use of carefully worded yes / no questions or multiple-choice questions, as well as the reducing of political dialogue to button-pushing, make a mockery of the democratic process at the same time that such a system hands over even more power to the technocrats.

But such a future is entirely compatible with the definition of "democracy" as merely a set of romantic preferences. Given the current zeitgeist of fascination with technology, interactive cable-TV would probably seem to many people to be the most exciting way to engage in political endeavour. What such a viewpoint overlooks is the process pinpointed by adman and political advisor Tony Schwartz: "The goal of a media advisor is to tie up the voter and deliver him to the candidate. So it is really the *voter* who is packaged by the media, not the candidate."[13]

Nevertheless, I suspect that the first major market for a new electronic TV-polling technology will be Toronto, the market-researcher's dream. No doubt those two thousand newly monitored Nielsen families would be thrilled to participate, having led the way into our bright new audimetered future.

Part III
The Global Agenda

10

The Global Pillage

The symbiotic growth of American television and global enterprise
has made them so interrelated that they cannot be thought of as separate. They are essentially the same phenomenon.

ERIK BARNOUW, *The Sponsor*

WHILE THE NEW medium's dramatic changes in the mid-1950s – from live
broadcasting to filmed product, from in-house network-made programming to a reliance on Hollywood film studios for filmed series, from network budgeting for in-house production to the deficit financing arrangement with producing companies outside the networks – had profound
effects on the American television scene, they had even greater repercussions world-wide. The switch from live broadcasting to filmed product
meant that American TV programs were now fully exportable. But even
more important, the economic change to a deficit financing base for production meant that U.S. export was virtually a built-in necessity.

While the producing company plays its waiting-game until the network
has run its licensed series twice, the company is seemingly losing money
until it can sell its in-the-can episode package as syndicated re-reruns on
the hundreds of local U.S. TV stations. But sales *outside* the domestic U.S.
market do not have to wait until the series has finished its two network
plays.[1] Such foreign sales are pure profit, and these sales financially carry
the producing company through the period of deficit financing until it can

begin to reap the bonanza of domestic syndication. Profits from foreign sales also continue to pour in long after a U.S. series has disappeared from domestic syndication.

The changes during the first decade of American television thus coincided with the postwar expansion of U.S. business around the globe. Erik Barnouw writes: "The power of the atom ended World War II. For all its horror, it was widely welcomed for the peace it brought. And it put the United States, almost unscarred by the war, in a position of extraordinary world-wide prestige and power."[2]

That position was enhanced by the Marshall Plan – ostensibly a program of foreign aid which, in fact, functioned as a means for granting export subsidies to U.S. companies already wealthy from lucrative war contracts. The "permanent war economy" advocated by the military and industry in postwar America necessitated the creation of global markets accesible to the permanently mobilized American business sector. Foreign aid after the Marshall Plan became "tied aid", meaning that the funds or credits had to be used to purchase U.S. goods and services. This ensured that recipient nations would be ready markets for U.S. multinational suppliers. In addition, countries accepting this tied aid were also often required to accept military aid, in the form of U.S. weapons, parts, and services. Thus, according to Erik Barnouw, U.S. foreign aid in the late 1940s and early 1950s "became a program by which the American taxpayer subsidized the global reach of the multinationals and at the same time fostered their close relationship with military hegemony."[3]

This situation was especially beneficial to the U.S. nuclear industry. Eisenhower's mid-1950s UN speech endorsing the "peaceful atom" launched a vigorous campaign to sell the wonders of nuclear power not just throughout North America but across the globe. General Electric and Westinghouse, at the time leaders in TV-set manufacturing, were also leaders in building nuclear-power generators for commercial use. Their twinned technological interests interesected first at the level of radiating devices but later at the level of globalized technological systems to be erected in the name of progress. Throughout the 1950s and early 1960s, dozens of countries were convinced to "go nuclear" in order to meet energy needs. The emerging Nuclear Club coalesced around a familiar ideology, expressed by Fred Knelman in his book, *Nuclear Energy: The Unforgiving Technology*:

> Nuclear power, the last hurrah of growth and technology, is an expression of a unified ideology of progress – resting directly on the omnipotence of technology, or the casual faith of technological optimism. For all true believers, nuclear power is both salvation and challenge. It fulfills and reinforces exist-

ing power structures.... It is the perfect toy for the technological personality, while it bureaucratizes and centralizes power for the powerful. It thus reinforces manipulative, elitist, and aggressive psychological, political and economic structures.[4]

By the mid-1950s, when the U.S. television industry had found its ideal production mode in an exportable, filmed TV product, and had established an economic structure that necessitated such export, American TV programming could become a significant factor in the larger global enterprise. As Andrew R. Horowitz observes, "Like the missionary campaigns of the 18th and 19th centuries, America's exported TV culture has served as a herald of empire. Its message has helped grease the international proliferation of U.S. business interests since the early Fifties."[5] As a "herald of empire", exported American TV would also express that "unified ideology of progress" resting on the assumed omnipotence of technology and the "casual faith of technological optimism". It would thus be central to the globalized technological agenda of U.S. expansion. But first the world had to be wired up.

□

The parent companies of the U.S. TV networks busily went to work overseas in the mid-1950s, selling the gospel of installing U.S. television hardware. U.S. corporations had patented and adopted the 525-line VHF system, in competition with higher-definition TV systems developed in France and Britain, which were actually superior technologically and yielded a better TV image. Since the competing systems were, in those days, technologically incompatible, U.S. business wanted the rest of the non-European world to adopt its chosen TV technology.

This sales job was no doubt eased by the power and prestige of the United States, the fact of "tied aid", and the presence of U.S. military bases dotting the globe. NBC's parent company, RCA – the leader in the electronics industry – sold entire television systems to foreign countries and then also supplied U.S. management expertise and technical aid. Ironically enough, RCA had begun its aggressive overseas enterprise in 1939 with the sale of a TV transmitter to the Soviet Union.[6] But the corporation's postwar sales to other countries were obviously fuelled by Cold War politics.

General David Sarnoff, head of *RCA,* accompanied his corporation's sale of TV hardware abroad with speaking engagements throughout the decade, in which he denounced the evils of communism.[7] Obviously, television was intended to be a linchpin in this offensive. Throughout the decade, RCA / NBC helped Canada, Portugal, Peru, Sweden, and Yugoslavia to set up TV systems, and went on to build stations in Egypt, Argentina, and

Hong Kong; a second TV network in Italy; and national TV systems in Kenya, Sierra Leone, the Sudan, Uganda, and Nigeria.[8]

Significantly, at this point historically the U.S.-Soviet space race was also on with a vengeance. Though the Soviets seemed ahead in the race towards the end of the 1950s, with their rockets sailing up into space while U.S. rockets were exploding on the launchpad, U.S. enterprise was well ahead on the ground. By 1958, the U.S. advertising agency Foote, Cone and Belding had conducted a study to determine the extent to which foreign markets were accessible to its American clients. The agency found that commercial television following the U.S. model for broadcasting was in operation in twenty-six countries.[9] Largely the work of RCA / NBC, this expanding television system continued through the 1960s. In 1964, the U.S. State Department, the U.S. Army Corps of Engineers, and RCA / NBC combined forces to build the largest television project ever undertaken by a U.S. firm: the construction of a thirteen-station network throughout Saudi Arabia. After this project was finished in 1966, RCA / NBC went on to build, again with U.S. government assistance, the national TV system for South Vietnam – to which RCA / NBC supplied (in the midst of the Vietnam war) technical advisors to supervise broadcasting.[10]

The involvement of the U.S. government and military indicates the fully ideological role that televison was and is intended to serve. As the herald of American empire, it would do more than ease the way for U.S. multinational enterprise. It would enfold the recipient countries fully within the U.S. sphere of influence. U.S. "advisory support" and management expertise provided to the U.S.-built television systems abroad meant that these technological systems were fully incorporated into the American agenda. Not surprisingly, the parent companies for CBS and ABC quickly followed the RCA / NBC example.

In 1960 ABC invested in television stations in Costa Rica, Honduras, Guatemala, El Salvador, and Nicaragua. As Andrew R. Horowitz details:

> The company then organized these stations into the Central American Television Network, to which it offered three important services: program buying, sales representation, and networking. The plan called for each station to relinquish its prime-time evening viewing hours to ABC, which, in turn, would supply it with "free" programming sponsored by American companies operating in the Central American market.[11]

In addition to these five Central American countries, U.S. television interests had, by 1964, become involved in controlling the TV schedules of twelve other Latin American nations: Brazil, Argentina, Mexico, Venezuela, Colombia, Chile, Peru, Uruguay, Panama, The Dominican

Republic, Ecuador, and Haiti.[12] This involvement clearly served to protect the holdings of U.S. multinational corporations throughout those countries, as well as to control the kind and amount of "information" that their populaces would receive. ABC also built Ecuador's first TV station in 1960 and helped establish the Philippine Republic Broadcasting System and the Arab Middle East Television Network comprised of stations in Syria, Lebanon, Kuwait, Iraq, and Jordan. Within a few years, ABC was maintaining television holdings in Japan, Australia, the Philippines, and at least fifty-four other TV stations around the world: to which it was supplying "assistance" in administration, sales, and programming.[13] CBS's foreign involvement, although nowhere near as extensive as that of RCA / NBC or ABC, included the construction of the Italian Broadcasting Corporation (RAI) in 1961 and the building of Israel's national TV system in 1966.

These expansionist policies of the U.S. TV-network parent companies coincided with a great global expansion on the part of the largest U.S. electronics firms during the same period. Benefiting from lucrative government contracts, particularly from the Department of Defense, the biggest electronics corporations – IBM, Honeywell, Xerox, Westinghouse, General Electric, RCA – had, by 1965, established over one thousand subsidiaries around the world.[14] Combined with the global agenda of the U.S. nuclear industry, U.S. enterprise throughout this period was definitely transforming the world into an electronic and nuclear interlocked system controlled by U.S. conglomerates.

In the midst of all this postwar American enterprise, in October 1962 Robert Sarnoff, Chairman of the Board for NBC, made a speech to the European Broadcasting Union, which had gathered in New York City. He told them:

> We are on the point of winning a gift that history has seldom yielded and never on such a scale. For centuries men have dreamed of a universal language to bridge the linguistic gap between nations.... Man will find his true universal language in television, which combines the incomparable eloquence of the moving image, instantly transmitted, with the flexibility of ready adaptation to all tongues. It speaks to all nations and, in a world where millions are still illiterate or semi-literate, it speaks clearly to all people. Through this eloquent and pervasive universal language, let us strive to see, in the words inscribed over the portals of the BBC, that "Nation Shall Speak Peace Unto Nation".[15]

The levels of mystification (to put it politely) operating in this speech are numerous, especially given the historical context. Just a few months before Sarnoff's address, *Saturday Review* had run an item that provided the real subtext for his remarks: "American television, having saturated time and priced itself to the top in the U.S., is on the way to conquering the Western Atlantic nations and Africa. Tomorrow, no doubt, the world."[16] The Sarnoff statement has a familiar ring to it, however, when we recall the McLuhanisms of "the wired nation" and "the global village" that were circulating at the time of the speech. These notions completely ignored the economic underpinnings of U.S. television enterprise and thereby played directly into its global agenda, giving techno-imperialism an aura of "natural" and neutral evolutionary principle.

The pattern regarding technological systems in general, and television in particular, is that organizations or nations are persuaded to invest in the hardware (a sign of "progress") without much thought given to the specific software needs or uses that accompany the investment in the technology. Throughout the 1950s and 1960s, most nations in the world were convinced by American corporate interests (as well as the U.S. government and military) to not only adopt American broadcasting hardware, but also to accept the commercial model for television broadcasting, regardless of whether or not that model (or TV itself) was suited to the particular country's needs. Once these stations had been established it became apparent that there was no local program-production base in place to fill the broadcast schedules. Moreover, such countries quickly realized that the cost of producing their own indigenous TV shows far exceeded the cost of readily available American TV product, which was "dumped" on foreign stations at cut-rates.

Here all those American advisors and managers provided to "assist" in foreign television operations played a most useful role. Economically, it made no sense for those operations to spend several thousand dollars in producing their own programs when, for a few hundred dollars, they could buy "dumped" American shows. This economic "logic", which was directly built into the commerical model for broadcasting exported by the United States, entirely served the needs of the American domestic industry under the dictates of the deficit financing structure. By exporting the hardware and the American commercial model, the U.S. television industry paved the way for its own "dumped" TV product and its national ideology to dominate the screens of the world. In some Latin American countries, where TV systems were built and programmed by U.S. interests, as much as 85 per cent of the broadcasting day was given over to American reruns.

In Canada, the politicians and businessmen of the late 1940s slavishly followed the lead of RCA / NBC advisers: adopting the U.S. 525-line TV technology so that viewers could tune directly to U.S. stations, copying the advertiser-based economic structure for Canadian broadcasting, incorporating the commercial goals for broadcasting despite a "public" network, and fostering private TV enterprises economically dependent on imported American shows.[17] Not surprisingly, TV broadcasting in Canada became overwhelmingly American in content, as much as in any Third World country.

Understandably then, given this history, United Nations researchers investigating the global flow of TV programs in 1974 found an extraordinary imbalance. The vast majority of countries with television systems were program-importers, making few shows themselves. Of the four major program-exporters in 1974, the United States was annually exporting 150,000 hours of TV product; the United Kingdom – 20,000 hours; France – 20,000 hours; and West Germany – 6,000 hours.[18] In other words, the United States was annually exporting three times as much TV product as the other three countries combined, and enough programming material to completely fill the broadcasting schedules of twenty-two networks operating eighteen hours a day for an entire year. With dumped American programming working to hamper the development of foreign program production, there were, by 1976, only five countries in the world whose television systems broadcast no American shows: mainland China, North Korea, Albania, North Vietnam, and Mongolia.[19] China, for one, has since changed its policy.

But one statistic puts all the others in perspective. As of 1977, the American domestic TV networks themselves were *importing* less than 2 per cent of their annual broadcast schedules – virtually excluding any and all television programming made in the rest of the world.[20] While American programming supposedly speaks a "universal language" (to use Sarnoff's phrase), foreign-made television is obviously reeking with difference.

Thus, the result of American global enterprise throughout the 1950s and early 1960s was that a particular kind of "global village" was created: one with the unmistakeable stamp of American origin and design. The Unyted States continually justifies its hegemony by upholding the notion of "the free flow of information" as a necessary democratic principle. This rhetorical notion overlooks the dictates of the economic structure for television put in place throughout the world by U.S. interests during the 1950s. It also conveniently ignores a particular political practice that is a highly significant factor in American domination of the world's screens.

□

Here we must return again to that historic change from live broadcasting to film. By turning the domestic medium over to the dictates of the Hollywood film industry, American television product would ever afterwards benefit from the aggressive practices of one of the most powerful lobby groups in the world, the Motion Picture Association of America.

The world's one billion TV viewers who, in 1986, dutifully tuned in to the annual rites of spring known as the Academy Awards ceremony, were able to witness Jack "Boom Boom" Valenti (this is how host Robin Williams introduced him) presenting the award for best foreign film. The irony of that particular conjunction between award-presenter and award-category was undoubtedly missed by most viewers, but outside the United States it should have been seen as the most blatant sign of American arrogance possible. For if there is any single person who has, for the past twenty years, done his utmost to actually *prevent* other countries from making their own feature films or even having their own indigenous film and television industries, it is Jack Valenti himself: former top advisor to President Lyndon Johnson and long-reigning president of the Motion Picture Association of America.

The sixty-four-year history of the MPAA, and the shorter but more lethal twenty-six-year history of its foreign division – the Motion Picture Export Association of America (MPEAA) – comprise a ruthless tale of strong-arm tactics and behind-the-scenes intervention in the domestic politics of most nations around the globe to prevent other countries from in any way interfering with U.S. domination over the movie and TV screens of the planet.

The MPAA / MPEAA represent the domestic and global interests of the top U.S. entertainment conglomerates: Columbia Pictures (owned by Coca-Cola), MGM / United Artists, Paramount Pictures (owned by Gulf and Western), Twentieth Century-Fox, Universal (owned by the Music Corporation of America – MCA), Warner Brothers (owned by Warner Communications), Buena Vista International (distribution arm for Walt Disney), Embassy Pictures, Tri Star Pictures, and Orion Pictures Corporation. These companies, through their subsidiaries, account for the lion's share of all American filmmaking and distribution, television series production, magazine and book publishing, music recording, and record / tape distribution in the United States and, increasingly, throughout the world.

Over the past twenty years within the United States, as mergers, acquisitions, and takeovers have crystallized media ownership in the hands of fewer than fifty corporations, the numerical presence of these media corporations on the Fortune 500 list has tripled. That list itself – the 500 largest U.S. corporations – comprises fewer than 1 per cent of the 360,000 U.S. incorporated businesses. But the Fortune 500 account for 80 per cent of all sales. "They are the aristocrats of the American industrial economy," writes media critic Ben Bagdikian, while the other 359,500 corporations are "the peasantry".[21] The largest media conglomerates, and especially the companies represented by the MPAA / MPEAA, "are now part of this American economic aristocracy". Not only do they dominate the U.S. domestic market, they earn more than 50 per cent of world-wide film profits, account for over 75 per cent of all TV programs imported around the world, and supply more than 50 per cent of all records and tapes sold across the planet.

Simply in terms of theatrical film rentals alone, the world-wide domination by these corporations (collectively known as the Hollywood "majors") annually generates a foreign box-office bonanza of more than $2 billion. (Canada contributes up to $400 million a year, the largest percentage from any one country, just from its movie theatres.) When we add television-program export revenues on top of that (about $500 million annually world-wide, $150 million from Canada), as well as video-cassette sales and rentals ($135 million annually just from Canada), we're talking about an annual global screen-bonanza of over $3 billion pouring into Hollywood from around the world.

But the export of American TV programs and films does more than merely generate 40 per cent of Hollywood's overall profit. The exported TV shows and feature films serve to pave the way for American enterprise, attitudes, and consumer products around the globe. The huge parent companies for the Hollywood majors – especially Coca-Cola, Gulf and Western, MCA, Warner Communications – are fully diversified into every aspect of communications, leisure, and entertainment, and they depend on exported American screen product to disseminate the American way of life from Burnaby to Bangladesh.

Thus, these giant conglomerates must ensure that foreign markets for U.S. screen material remain as open and accessible to the majors as possible, without any "unfair" trade restrictions to hamper that accessibility. Their key lobby – the MPAA / MPEAA – is committed to maintaining that accessibility. In 1960, with television operations expanding around the globe through the "assistance" of U.S. TV networks, the State Department, and the military, the MPAA cloned itself to create the MPEAA and vested it with the authority to represent its members in negotiations with foreign governments, networks, and entrepreneurs. Although the U.S. screen industry had already ensured that most foreign television systems were set up according to its technological and commercial model and would rely on "dumped" American TV series, it was useful to have an agency representing its interests in an active way.

For any country to have a film industry, it must have local control over three aspects vital to successful operation: production, distribution, and exhibition. It does no good for a country to pour millions of dollars into film production if distribution and exhibition are controlled by alien interests. But that is precisely the role of the MPAA / MPEAA world-wide: to ensure that the distribution / exhibition aspects of the industry benefit only the movies and TV programs made by its members. Here it's important to note that most nations, following the coaching of American enterprise, slavishly abandoned live TV broadcasting (if they had, indeed, ever adopted it) and made the shift to filmed product. Obviously, then, the health of their own film industries would be crucial to the fate of their television systems. And it was precisely there that the MPEAA would play the decisive role.

While the MPAA looks after U.S. domestic matters, the MPEAA operates as a single bargaining unit with foreign customers. This activity is prohibited in the United States itself on anti-trust grounds, but it flourishes abroad under the protection of the Webb-Pomerene Act of 1918, which permits U.S. businesses overseas to function as monopolies with a single sales agent empowered to set prices and arrange contracts for all its members' products. In its foreign activities, the MPEAA is not unlike the U.S. State Department, with which it has ties, and one of its quasi-governmental functions has been to lobby against any foreign legislation that would restrict the activity of the Hollywood majors. Through the U.S. State Department, the MPEAA can threaten boycotts and trade reprisals for countries that do not co-operate with its interests.[22]

Despite such practices by the MPEAA since its founding in 1960, several countries have enacted legislation over the years to regain some control over their TV and movie screens. Often, such legislation has involved setting quotas that dictate a limit on foreign screen product, with a certain percentage of screen-time given over to indigenous productions in order to foster that country's own television and film industries. Britain has had one of the more stringent quotas on foreign TV product. In 1978 its limit on foreign TV imports was a stiff 14 per cent of the schedule. In that year, its regulatory agency for broadcasting, the Independent Broadcasting Authority, planned to tighten that quota even further to 12 per cent, thereby freeing more broadcast time for home-grown programs and fostering employment within its own production industries. The MPEAA moved in quickly to fight the plan, hinting at "retaliatory action" that eventually forced the IBA to instead raise its foreign quota to 15.5 per cent.[23] In that same year, Spain balked at the latest price increases proposed for "dumped" U.S. programs in that country. After a boycott threat by the MPEAA, Spain accepted a price mark-up of 100 per cent for such product. *Variety*, the voice of the U.S. industry, called the MPEAA threat "part of a larger effort to land higher prices for U.S. films and TV shows in various countries in Europe and the Middle East".[24]

More recently, in Quebec, legislation was on the verge of being enacted in 1985 to restrict the activities of the Hollywood majors, who control and dominate the screen industries of not just Quebec, but Canada at large. Direct pressure from the U.S. State Department and the MPAA on the Quebec Cabinet, combined with a threat to boycott the province, resulted in a

Cabinet cave-in on the impending legislation.[25] So much for the Sarnoff rhetoric about "nations speaking peace unto other nations".

Ever since the MPAA / MPEAA recognized the extraordinarily lucrative bonanza in exporting screen product throughout the world, it has aggressively lobbied against what it considers to be the top ten threats to its members' domination of the world's TV and movie screens. Those "unfair trade practices" adopted by other countries to foster their own industries include: (1) import quotas, (2) screen-time quotas, (3) discriminatory admission taxes, (4) film rental controls, (5) currency remittance restrictions, (6) requirements that prints be struck in the country of origin, (7) dubbing requirements, (8) foreign prohibition against alien distributors, (9) high income taxes, and (10) production subsidies / incentives. Whenever any country has attempted to legislate such measures, the MPAA / MPEAA has moved into high gear, with the help of the U.S. State Department, to block foreign efforts to develop their own indigenous industries. The "global village" constituted through the TV and movie screen of the world must remain under U.S. control and exploitation.

Both the Soviets and the Americans have ringed the world with nuclear weapons. But the United States has also accompanied its nuclear domination with colonizing practices that fully enfold most of the world into its spheres of influence. Consumption, whether in the form of products or images, is obviously the way to ensure that U.S. multinational hegemony remains unchallenged. With the mainland Chinese watching Disney TV shows and drinking Coca-Cola, and the Soviets watching *Dallas* and drinking Pepsi, perhaps we must now view the world in terms of "colawars", not the Cold War.

Nevertheless, historically the spread of U.S. television hardware paralleled the spread of its nuclear hegemony and military-industrial might. In many instances these two systems overlapped on the TV screens of the world in an overt way. From the mid-1950s and throughout the 1960s, pro-nuclear films made for the U.S. Atomic Energy Commission or for companies such as General Electric were widely distributed on those foreign TV systems set up by the American networks. Thus, millions of viewers around the world were treated to films like *The Atom and Eve* (in which nuclear power was pitched to women as the best way to provide electricity for their domestic appliances), and *A Is For Atom, Return To Bikini, Go Fission, No Greater Challenge, The Mighty Atom* – all part of the AEC's public-relations activities to sell the nuclear industry abroad.[26]

□

By controlling and dominating the screens of the world, the United States insists quite literally that its "vision" is the only one that shall prevail. Perhaps even more powerfully than its military supremacy, this techno-imperialism over the airwaves and the imaginations of the world's population has caused, and still causes, extraordinary rifts and tensions as nations struggle against the overwhelming embrace of the global American agenda.

Should we be surprised, for example, that in many of those countries where U.S. interests erected and controlled television as the technological herald of empire, there now erupt fierce struggles for nationhood and independence? And similarly, should we be surprised that so-called "terrorist" acts are specifically staged with television coverage in mind? Since the American TV-screen shows virtually nothing from the rest of the world, while simultaneously deluging other countries with its own product, that U.S. TV-screen itself becomes the target for conveying, through even momentary news coverage, situations and conditions otherwise excluded from the American frame of reference.

It is the particular effectiveness and efficiency of the medium of television that, quite literally seeing through it, we tend not to see *it* and include it as in any way a highly significant contributor to problems. Moreso perhaps than any other electronic medium, television easily plays a disappearing act: framing everything else while vanishing as the frame. As we become more conscious of it as the frame – and especially as the frame erected, provided, and maintained world-wide by specific U.S. interests – we begin to grasp its central role in postwar history. In particular, we need to bear in mind the words of Octavio Paz, which almost fully summarize television in both its roles as "herald of empire" and spectacle endorsing the "omnipotence of technology":

What sets worlds in motion is the interplay of differences, their attractions and repulsions. Life is plurality, death is uniformity. By suppressing differences and peculiarities, by eliminating different civilizations and cultures, progress weakens life and favors death. The ideal of a single civilization for everyone, implicit in the cult of progress and technique, impoverishes and mutilates us. Every view of the world that becomes extinct, every culture that disappears, diminishes a possibility of life.[27]

11

Nuclear TV

THE GREATEST PROBLEMS of our times are largely boundary issues. This is apparent not only in the political strife around territorial disputes, but also in the increasing threat to the planet from phenomena that bypass barriers of every sort. Acid rain, air and water pollution, radiation leaks, nuclear fallout, the impending threat of nuclear war itself – all bypass both geopolitical borders and the personal boundaries of the individual. The technologies and practices of our time have placed the very notion of boundaries at risk. "Good fences make good neighbours," wrote the poet representative of an earlier era. While the wisdom in the line is still appropriate and even crucial to survival, it has become extraordinarily difficult to know what materials might begin to make the proper fence.

Television, as technology and practice, is amazingly efficient at eliminating borders and boundaries. Technically, its signal transmission through the airwaves makes it like any other form of invisible particle carried across geographical distance and political borders, though obviously, specific business practices and technological advances (such as satellites and cable) have greatly enhanced its reach. Similarly, television bypasses personal boundaries, especially the boundaries of self / screen, conscious / unconscious, and even technology / body in the sense that the nervous system responds to the electronic scanning process.

Another aspect of the medium's capacity to eliminate boundaries came to light in the middle and late-1960s, when a series of events exposed the

TV receiver-set to be a device emitting significant levels of ionizing low-level radiation into the homes of millions of viewers. In retrospect, the sequence of events reveals not only how sacrosanct the television industry is, but how necessary the fiction surrounding "safe" levels of radiation is to our entire way of life. As Rosalie Bertell has stated: "There's been a campaign since 1951 to convince the public that low-level radiation is harmless.... People have a right to know what's happening to human health."[1]

Throughout the 1950s television played a crucial part in that campaign through various aspects of its role as spectacle. But the revelation that the medium itself was emitting radiation in a very bomb-like manner was something else again. The subsequent handling of the issue reminds me of the words of Willard Libby, the Atomic Energy Commission's fallout specialist in the early 1950s: "People have got to learn to live with the facts of life, and part of the facts of life are fallout."[2]

□

The TV-radiation issue that exploded in the late 1960s was far more than a mere tempest in a teapot. In the days before solid-state components, television sets were discovered to be emitting levels of radiation up to one hundred thousand times the recommended "safe" level of 0.5 millirems per hour.[3] When a skeptical scientific team of the U.S. Bureau of Radiological Health (BRH) turned its attention to the issue, its findings were a complete surprise even to team members themselves, as the Bureau's Dr. H.D. Youmans admitted:

> "We questioned whether TV radiation was important, because it was so low compared to the output of an x-ray machine," Youmans said. "We thought the rays would be soft and nonpenetrating. Instead, we found rays escaping from the vacuum tubes to be harder and of higher average energy than we expected. They penetrated the first few inches of the body as deeply as 100-kilovolt diagnostic x-rays. You get a uniform dose to the eye, testes and bone marrow."[4]

Another member of the BRH scientific team, Dr. Norman Telles, revealed that his group had initially assumed that there was a "safe" threshold below which low-level radiation ceased to penetrate the body. But following their research he concluded, "We have made the assumption that there is no threshold, that radiation down to the zero level evokes a response from body tissues."[5]

Since 1946, television receiver-set manufacturers had been aware that their product emitted low-level radiation.[6] This radiation is the result of the electronic principle by which high-voltage vacuum tubes work. In simplest terms, the TV screen is composed of phosphors that glow when electrons collide with them. As an electron is fired by the cathode-ray tube, it is accelerated to a very high velocity and has greatly increased energy. When it strikes the picture tube, the electron's velocity is reduced to zero and, therefore, it must give up its energy. Part of that energy is converted to visible light by the phosphors illuminating the screen. Another small part is converted to heat which is absorbed by the glass, while most of the remaining energy is converted into x-radiation. *Time* magazine, in the heat of the controversy, explained the electronic principle this way:

> Any vacuum tube operating at several thousand volts or more produces detectable x-rays. Boiling off the incandescent cathode of the tube, electrons are attracted and accelerated by the high positive voltage on the tube's anode and smash into it at great speed. Struck by the electrons, the atoms of the metallic anode vibrate violently and emit energy in the form of x-rays, which can burn the skin, injure the eyes, and cause genetic damage. To reduce x-ray emissions of high voltage tubes to safe levels, manufacturers [of TV sets] equip the tubes with metallic shields that absorb most of the radiation.[7]

In the pre-solid-state television receiver there were several high-voltage tubes which were eventually revealed to be the source of radiation: the cathode-ray picture tube, the high-voltage regulator tube, and the high-voltage rectifier tube.[8]

As early as 1948, the Victoreen Company in the United States had invented a diode that could have replaced the high-voltage shunt regulator tube, thereby eliminating one of the sources of radiation in TV sets. Apparently, because the levels of radiation emitted by black-and-white sets were not considered significant enough to call for this alteration in set design, the industry did not incorporate the Victoreen diode. The military, however, decided to use it in other devices based on vacuum tubes: radar display circuits and geiger tube circuits, for instance.[9] With the introduction of colour TV sets, the industry faced increased levels of radiation emissions from sets because the colour receiver works at a much higher voltage than the black-and-white set, and depends on three cathode-ray guns in the picture tube to produce the colour spectrum, as opposed to one for black-and-white. Despite this prospect, manufacturing costs appeared to be the deciding factor. The Victoreen diode "was used by three manufacturers of early color television receivers, but cost prohibited its consideration when the market became highly competitive."[10]

From 1946 until 1962 there is no published indication that I could find to suggest that the public was in any way informed of the possibility of low-level radiation emissions from their television sets. Then, in 1962, a writer for *Popular Mechanics* briefly speculated on the question that would erupt five years later into the public consciousness. In a short piece entitled, "Is Your TV A Radiation Hazard?", Dr. Jan Paul noted, "The x-ray tube and the TV tube are both radiation-emitting cathode-ray types." But he quickly concluded, "If there is any hazard from viewing TV, parents will have to blame it on programming rather than the radiation, which is insignificant."[11]

Then, in November 1964, an independent scientist named John N. Ott came across a report, presented to the American Academy of Pediatrics, which detailed a study of thirty children suffering symptoms of nervousness, continuous fatigue, headaches, loss of sleep, and vomiting. Since the common factor among these children seemed to be that each was watching between twenty-seven and fifty hours of TV every week, doctors had prescribed a total abstinence from TV-viewing. According to the report, in those cases where parents enforced the prescription, the symptoms vanished in two or three weeks. Where the parents relaxed on enforcing this abstinence, allowing their children to resume their usual amount of TV viewing, the symptoms returned.[12]

Intrigued by this report, which offered no concrete explanation for the symptoms, Ott decided to try to determine if there was any form of radiation emission that might account for the children's reactions. For his experiments, conducted in 1964 and 1965, he used a large-screen colour TV, and covered one-half of the picture tube with one-sixteenth-inch solid lead – the protection customarily used to shield x-rays. He covered the other half of the picture tube with ordinary heavy black photographic paper – the kind that stops visible light but allows other radiation to penetrate.[13]

Ott's first experiment involved putting this TV set in a greenhouse and placing potted bean seeds in front of the TV screen: six at various levels in front of each half of the screen, and another six fifty feet away from the greenhouse. After three weeks Ott observed that the seedlings in front of the lead shielding, and those located fifty feet away, displayed a normal growth of six inches. The plants in front of the paper-covered side of the TV screen had leaves three times the normal size and a growth of up to thirty-one and a half inches. Plants whose pots had been placed in line with the top of the TV set all had roots that grew upward, emerging from the top of the soil. Ott writes:

I later learned, in talks with scientists at the United States Aerospace Medical Center, that wheat seedlings, orbited in a biospace capsule, had behaved in a strikingly similar manner. The random growth of the wheat was thought to be due to weightlessness.... A more logical explanation for the wheat might lie in the fact that the space capsule was being bombarded from all directions by radiation, as were the beans in the pots.[14]

Ott decided to put white laboratory rats in front of the TV set. "The rats protected only with the black paper became increasingly hyperactive and aggressive within from three to ten days, and then became progressively lethargic. At 30 days they were extremely lethargic and it was necessary to push them to make them move about the cage."[15] He also introduced a second TV set (black-and-white), located at the so-called "safe" distance: six feet away from the cages. Within ten to twelve days, all the young rats in one of the cages died. Two others that were lethargic were taken to an animal pathology laboratory where they also soon died. Autopsies revealed what seemed to be brain damage.

Throughout these experiments, Ott made time-lapse films that showed the changes in both the seedlings and the rats. In 1965, he screened these films for research-engineering teams at two large (unidentified) TV set manufacturing companies. One of the companies refused further communication with him, while the other prompted a letter, dated August 6, 1965, from the Electronic Industries Association: the chief lobby for the entire industry. The letter assured Ott that all TV sets made in the United States met the current recommended radiation-emission standard of 0.5 millirems per hour when measured at five centimeters from the surface of the TV set. Reminding Ott that this recommended standard had been determined by the National Committee on Radiation Protection and the International Commission on Radiation Protection, the letter went on to state: "At this level, no detectable somatic injuries were expected, even if the level were to be exceeded by a factor of 100."[16]

What is interesting in this statement is the use of the phrase "detectable somatic injuries", as though low-level radiation does not accumulate in the bone marrow and cells of the body over time, without leaving a visible trace on the surface of the skin. As well, the statement indicates (as Ott himself notes) that the Electronic Industries Association in 1965 considered fifty millirems per hour of low-level radiation (0.5 mr / hr. "exceeded by a factor of 100") to be a "safe" level of exposure. To put this figure into a context, we can consider a statement by Dr. Rosalie Bertell:

> I had been measuring the health effects of one, two, three, four, and five chest x-rays. Then I found that the federal government [in both Canada and the U.S.] allows the general public to receive up to five hundred millirems per year. That is equivalent in bone marrow dose to one hundred chest x-rays per year. That was really shocking![17]

Fifty millirems per hour – the emission level apparently considered "safe" by the set-manufacturing lobby in 1965 – would be the equivalent of ten chest x-rays carried out in an hour's time. Even granting that low-level radiation diminishes over distance travelled from its source, the figure is remarkable as a supposed limit for safety.

Shortly after Ott's attempts to alert TV manufacturers to his findings, RCA decided to conduct its own experimental study on bean seedlings and rats. Hiring an independent research lab, the Bio-Analytical Laboratory of Freehold, New Jersey, RCA subsequently reported that no abnormal biological effects had been found. RCA's director of research confidently told the press that the possibility of radiation from TV sets was "under complete control by the entire industry" and that sets were constantly checked by the Underwriters Laboratories for x-rays. "It is utterly impossible for any TV set today to give off any harmful x-rays," the RCA spokesman stated.[18] Complicating matters, however, the Underwriters Laboratories later revealed that their accepted emissions standard was 2.5 millirems per hour: five times higher than the recommended level.

Undaunted by such industry reactions to his work, Ott showed his time-lapse films at a meeting of the 100th Technical Conference of the Society of Motion Picture and Television Engineers on October 3, 1966. Although I could not find a single reference to TV radiation-emissions in the Society's SMPTE Journal from 1966 through 1972, the screening must have had an impact throughout the industry, as events just one month later suggest.

☐

In 1966 General Electric found itself in an unusual position economically. By that year, GE (along with Westinghouse) had a monopoly on the export of nuclear generators to the non-U.S. Western-bloc market.[19] On that front, business seemed to be booming. But in TV set sales GE figures for 1966 were low and the company was due to begin laying off 3,200 workers from its Major Television Department in December. One month before these layoffs were to begin, GE technicians discovered highly significant levels of ionizing radiation being emitted by its colour TV sets – levels later revealed to have been up to one hundred thousand times higher than the recommended standard of 0.5 millirems per hour. The response of GE to this discovery is worth noting, especially given that many thousands of these defective sets had already been sold.

Rather than alert the public to the hazard, for six months General Electric chose to maintain a strict silence about the situation. During this period the company tracked down the cause of the excessive radiation (an improperly aligned metal shield for the high-voltage shunt regulator tube), redesigned the set by February, and in April 1967 began the task of locating and repairing all the defective TV sets that had been sold – still without informing the public.

"We were certain," said a GE spokesman later, "that an announcement would develop unfounded fears and disrupt the field program [of finding and repairing the sets] with an inundation of phone calls".[20] Of course, a public announcement might also have affected the company's prospects for increasing its TV sales, already low by comparison with previous years' figures. Sometime during this six months of silence, GE did manage to report its product defect to the U.S. Public Health Service's National Center for Radiological Health (NCRH), but that august organization also decided to maintain a polite silence on the matter.

Meanwhile, 154,000 owners of defective GE televison sets were unknowingly receiving more than entertainment from their recent purchases. The defective sets emitted ionizing radiation at the level of fifty thousand millirems an hour (0.5 mr / hr. multiplied by 100,000), potentially exposing owners, in just one hour's viewing, to ten times the current yearly exposure standard applied to employees in the nuclear industry: five thousand millirems, or five rems per year.

Instead of blowing the whistle on General Electric, the NCRH began urging the company to make the necessary announcement to an unsuspecting public – a procedure obviously geared to trying to "save face" for the corporation. If the company made the announcement itself it would retain some semblance of corporate responsibility about the matter. Undoubtedly, both the National Center for Radiological Health and GE must also have sensed that such an announcement could open a huge can of worms on an issue that had been hidden beneath the surface since 1946.

With both NCRH and General Electric reluctant to break the silence surrounding the issue, the public remained blissfully ignorant of the situation from November 1966, when the defect was discovered, until early May 1967, when the story was somehow leaked to a *New York Times* reporter who began asking the usual impolite questions of a sensitive nature. This prompted GE to issue a press statement on May 18, 1967 announcing its field-program to "modify" ninety thousand large-screen (eighteen inches or larger) colour receivers that had been manufactured and sold since June 1966. As reported in *The New York Times*:

> General Electric said that "preliminary findings", confirmed by "nationally recognized radiological health experts," had found that the emissions were not "sufficient to cause harm to viewers."
>
> "However," General Electric said, "in view of its 26-year effort in providing customers with reliable and safe TV receivers, the company has a

program well under way to make appropriate changes in the models involved. These changes will resolve any problems that may exist and will materially improve performance and reliability."[21]

The GE statement made no mention of the levels of radiation issuing from its defective sets (the number of which later reached 154,000), simply saying that it had discovered "x-radiation in excess of desirable levels".

Reached for comment on the matter, James G. Terrill Jr., director of the National Center for Radiological Health in Rockville, Md. (the same agency which had neglected to blow the whistle on GE), issued a carefully worded statement that, in retrospect, can be seen to have nicely absolved the NCRH silence: "As of now, there is no evidence in the hands of the National Center to suggest that any TV receivers manufactured by the General Electric Company or that sets made by other companies have excessively exposed viewers of TV sets."[22]

Coincidentally, the GE announcement of its product defect occurred at the same time that a proposed piece of legislation was pending before U.S. Congress to create a national commission on product safety. Co-sponsored by Senator Warren G. Magnuson (D-Washington), chairman of the Senate Commerce Committee, and Norris Cotton (R-New Hampshire), the Bill proposed a seven-member commission to assess product hazards and determine whether current legislation was adequate. Within this legislative context, Congressman Paul Rogers (D-Florida) requested, a week after the GE announcement, that public hearings be scheduled to investigate the possible radiation hazard to TV viewers.

Given this context, it's worthwhile to consider how the radiation issue in GE's announcement was treated by some of the mainstream U.S. press: the first wave of information, which would colour subsequent treatment of the story. *The New York Times,* next to its page seventeen article on *GE,* which cited other set manufacturers as having "no radiation problems", printed a small item entitled "Health Service Advice". This item, focusing on advice from Mr. Terrill of the NCRH regarding GE set owners, mentions "weak x-rays", "low energy x-rays" emitted downward in a "thin" crescent pattern, yielding a radiation dose that would be "slight". According to the item:

> At a range of roughly one foot below the set, [Terrill] said, this would be about the size and shape of the crescent a man could make with his thumb and forefinger. At a greater range the crescent would be bigger, but the radiation weaker, he said. Mr. Terrill estimated that a person who sat directly under such a set for a cumulative total of 40 hours might receive enough radiation to any exposed portion of his body to make the skin red and somewhat painful.[23]

The language used here arguably has the effect of completely diminishing the issue, literally reducing it to dimensions that neatly fit between two fingers of a hand and couching it in terms that might even seem mildly amusing: the image of someone lying beneath his TV set for forty hours as his skin turns red. The article nicely avoids two significant facts: 1) the actual radiation emissions levels discovered from the GE sets (up to 50,000 mr / hr.); and 2) the cumulative effect in bone marrow, testes, and cells. Thus, the invisible danger is displaced by a visible effect (red skin), which is made to seem somewhat amusing.

A somewhat similar treatment occurred in the reportage of *Business Week* (May 27, 1967), which hastened to reassure readers that the level of emissions and the direction of the "small beam" of x-rays from the defective GE sets "make it unlikely that the maximum injury – a slight reddening of the skin – could occur."[24] *Science News* (July 1, 1967) seemed to take the problem more seriously, stating, "Defective sets contribute to the total genetic radiation dose of the population, as well as causing possible cornea damage, such as cataracts, to TV viewers". The article quoted Dr. Harold Stewart, director of the NCRH x-ray exposure laboratory, who offered warnings about "the possibility of slight mutations to persons exposed to a direct beam [of TV radiation] for a certain number of hours". Stewart, the article remarked, "was unable to be more precise".[25]

Like the "slight reddening of the skin" mentioned in the other press accounts, these "slight mutations" might well have seemed easily dismissible in the context of the times. *Consumer Reports* (July 1967) seemed to slight the issue further, telling its readers, "Don't worry about using your color set". It informed them that manufacturers "seem to have agreed to limit the amount of x-ray emission ... to what authorities consider to be an acceptably safe level".[26]

Not surprisingly, throughout the summer of 1967 the press had far more significant stories to cover, especially because the TV-radiation emissions issue appeared to involve only one company and a few thousand sets. Those items that did appear tended to emphasize a containment of the problem by focusing on the specifics of the GE case: the modification in the field of those GE sets with an eighteen-inch or larger picture tube, made between June 1966 and February 1967, with a K-C colour chassis and 6EF4 or 6LC6 regulator tubes, a serial number beginning with the letters OA or OD (or in some cases, no serial number at all), and a blue "fine tune" gauge above the control knob. Given these specifics, worried set-owners need only check out such details to be reassured that their TV sets must be "clean".

Nevertheless, by late summer of 1967, some larger issues had begun to emerge. *Newsweek* reported, "This week GE and the NCRH may be called upon for more details when the House Subcommittee on commerce and finance begins hearings to find out how a potential health hazard slipped through both the stringent quality controls of a major firm and the surveillance of a Federal agency."[27] Meanwhile, the U.S. Public Health Service (overseer of the NCRH) began a program of testing the radiation levels from colour TV sets made by various manufacturers.

As the House Subcommittee hearings got underway, the Electronic Industries Association began to move into high gear as the major lobby for a threatened industry. Association spokesmen appeared at the Subcommittee hearings in October 1967 to give wholehearted reassurances: "Investigation has disclosed that this [the GE incident] is an isolated case and is not representative of television set operation."[28] While this quietening view was quoted in the press, a less optimistic position on the issue was kept under congressional wraps. "We brought [John N. Ott] to Washington for a briefing," said Representative Paul Rogers later, "but we didn't say anything for a while. We were afraid we would scare a lot of people. And when we checked with the U.S. Surgeon General, he told us there was nothing to worry about."[29] Ott's findings suggested that the situation was no "isolated case" but rather one embracing the electronic medium itself.

□

Despite industry and government precautions, the issue was not one that would easily disappear. Even President Lyndon Johnson (himself a TV station owner) further alerted the public by calling for, in his State of the Union Message of January 1968, "protection against hazardous radiation from television sets and other electronic equipment" – a statement that must have induced a slight reddening of the skin on the faces of industry lobby spokesmen. Shortly after Johnson's message, the Public Health Service announced the results of its survey of 1,124 colour TV sets in and around the area of Washington, D.C. Excessive radiation (in some cases twenty-five times higher than the recommended level) was discovered coming from sixty-six sets: about 5 per cent of the total examined. These sets were from ten different manufacturers, leading the NCRH's James Terrill, Jr. to reverse his previous position and state, "The problem of excessive radiation is not limited to the sets of one manufacturer."[30] The U.S. Surgeon General issued a warning that TV viewers should sit from six to ten feet away from their sets, though he did not explain why this distance was considered "safe". Since the PHS findings indicated that radiation was most prevalent at the rear and sides of the TV set (not at the bottom, as was the case with GE), viewers were warned to avoid prolonged exposure to those areas.

The published material between 1967 and 1970 shows a subtle but concerted effort to privilege one part of the TV set as nonradiating. Over and over again, industry statements asserted that the picture-tube was not to blame for any radiation. This rationale was a part of the first proclamation from GE on May 18, 1967: "[GE] said the picture tube was not the cause". It was reiterated over the years until 1970, when a writer for *Radio-Electronics* (not a general interest magazine, obviously) included the picture-tube as one of the three sources of radiation from TV sets.

This subtle privileging of one part of the technology was undoubtedly connected to the place of the viewer. The sacrosanct viewing space in front of the screen could not be endangered without raising the possibility of a wholesale desertion of the place assigned to the consumer, not just in front of the set but, through it, as a member of a society of spectacle. John Ott's initial experiments implicated both this sacrosanct place, and the picture-tube, as the site of radiation effects. Given this it is not so surprising that the congressional subcommittee decided to remain silent about Ott's testimony at the hearings.

As the hearings proceeded throughout 1968, the industry lobby stopped trying to assert that GE's radiation problem had been an isolated case, and instead turned to pursuing another significant tack. *Consumer Reports* noted in its September 1968 issue:

> Unfortunately, the industry has ... been lobbying hard to weaken legislation that would establish Federal radiation standards for all electronic products and [that would] require strictly enforced controls, including factory inspection and mandatory callbacks of defective sets at the manufacturer's expense.... By mid-1968, there were 16 million American homes with color TV sets. If it could be assumed from the PHS survey that 5 per cent of those sets exceed the acceptable radiation limit, then some 800,000 households, consisting of 2.5 million people, are being needlessly exposed to some small amount of radiation.[31]

In the midst of the controversy, Westinghouse decided in late 1968 to stop manufacturing colour TVs, opting to distribute sets made elsewhere. This left the market to the "Big Three" that controlled roughly two-thirds of all sales: RCA, Zenith, and Magnavox. The remaining share was divided primarily among General Electric, Philco-Ford, Sylvania, and Admiral. Interestingly, none of these industry leaders took the first move into solid-state components to try to solve the radiation problem. Instead, a small company, Motorola, brought out the first solid-state colour set, in 1968.

During the last days of the 90th Congress, in October 1968, a piece of legislation was passed which set the stage for the establishment of safety

standards for electronic products. Up until that point, there had been recommended safety levels for radiation emissions adopted on a voluntary
basis. The Radiation Control for Health and Safety Act of 1968 required
that the Department of Health, Education and Welfare (HEW) set standards which would strictly limit permissible radiation from TV sets and
other electronic products by January 1, 1970. However, as a source of
authority for enforcement powers necessary to protect consumers, the Act
appeared to be less than ideal.

In the Senate, the Bill co-sponsored by Magnuson and Cotton had
empowered the government to seize substandard products and to inspect
factories when necessary. By the time the Bill had passed in the House, the
new Act withheld seizure powers and made no clear provision for factory
inspection. Behind closed congressional doors, J. Edward Day, lobbying
spokesman for the Electronic Industries Association, had effectively convinced members of the Senate-House Conference Committee to gut the
radiation protection Bill.[32] But as events continued to unfold, the need for
a strong Bill was becoming increasingly obvious.

In April 1969, the Public Health Service of Suffolk County, New York,
released the results of its two-year survey of five thousand colour sets. Its
report, in contrast to the earlier PHS survey in Washington, showed that 20
per cent of the sets (over one thousand receivers) were emitting radiation in
excess of the recommended standard of 0.5 mr / hr. Scientific reaction (at
least in official circles) to the differences in these two surveys tended to dismiss the more alarming report. As *Good Housekeeping* told its readers,
"Of the two surveys, Dr. Hanson Blatz, director of radiation control of the
New York City Health Department, called only the Washington study
'truly scientific', maintaining that the instruments used were more precise."[33] However, later research revealed that the different geographical
locations of the two surveys could yield such a discrepancy in findings:
because reception in outlying suburban areas is poorer than within a city,
set-repairmen and owners tended to boost the picture-tube voltage in
order to get a clearer picture. This increased voltage was found to be exas-
cerbating the radiation-containment problem, making the sets surveyed in
the Suffolk County outlying area four times more prone to excessive emissions than those in an urban milieu like Washington, D.C.

By June 1969, HEW had named a fifteen-member committee, equally
divided among representatives from the electronics industry, government,
and the public, to draft radiation standards for TV sets, microwave ovens,
and medical / dental x-ray equipment. Early on in its meetings, this committee received the submission from the Bureau of Radiological Health,
which took a strong position on the dangers of low-level radiation to

health. In its submitted draft of proposed manufacturing standards, the BRH recommended that the level of 0.5 mr / hr. be adopted until 1971, at which time a stricter standard of 0.1 mr / hr. (measured at a distance of five centimeters) should go into effect.

Several members of the HEW committee strongly objected to the Bureau's proposal, calling it too tough on the industry and stating that the suggested level of 0.1 mr / hr. would have to be reviewed to determine whether it was "reasonable and technically feasible". These members protested that there was not adequate instrumentation available to the industry to detect radiation down to that level.[34] When the HEW committee finally settled on the standard of 0.5 mr / hr. as "safe" for electronics products (rejecting the lower figure of 0.1 mr / hr.), it defended its decision by using the argument that had been advanced by some of its members. It was an argument nicely punctured by *Consumer Reports* in its January 1970 issue:

> The 0.5 mr / hr. level was picked, says HEW, because radiation levels under 0.5 mr / hr. cannot be measured reliably. CU [The Consumers Union] disagrees with that contention. In fact, according to published information from the Electronic Industries Association ... manufacturers have been routinely measuring levels below that.[35]

In response to HEW's decision consumer advocate Ralph Nader stated: "The standards are too low. Millions of people are being exposed to the risk of physical, genetic and eye damage.... The forces of industry and bureaucracy have prevailed."[36]

The new federal standards for radiation emissions from TV sets and other electronic products came into effect on January 1, 1970, in three stages. As of that date, all newly manufactured TV sets were to meet the level of 0.5 mr / hr. when properly adjusted. The second stage, in effect as of June 1, 1970, required that receivers manufactured after that date must remain within the stated limit of emissions even when the controls on the set were maladjusted in a way that could increase the radiation (as in boosting the voltage to get better reception). As of June 1, 1971, the standard of emissions would cover component or circuit breakdowns in the set as well.

In response to the new legislation, which did little more than make mandatory what had previously been the recommended emission standard, the TV manufacturing industry began moving in a variety of ways towards the production of cleaner sets, and eventually to the use of solid-state components that (apparently) eliminated the radiation problem. The spokesman for the Electronic Industries Association, J. Edward Day, remarked at the time, "These efforts are being pursued not because there is any feeling on the part of TV manufacturers that a hazard situation exists or that there is any justifiable cause for public alarm." It is rather, he said, an effort "to bring an end once and for all to the flurries of public excitement over TV radiation".[37]

□

As these flurries of concern died to an occasional murmur by the end of 1970, primarily because of the widespread sense that legislative action had solved the problem, one of the last pieces published on the issue might well stand for the mainstream position taken throughout the entire episode. Dr. Jan S. Paul, this time writing for the *Consumer Bulletin,* dismissed the danger of low-level radiation to various parts of the human body unless the dose was massive. Nevertheless, he wrote:

> In one area, that of red blood cell structure, there is reason for just concern, because it has been found, again under laboratory conditions, that red cells can be affected by relatively low levels of radiation. Fortunately the condition is almost always temporary, and total recovery comes fast upon removal of the source. Of course, therein lies the rub, because it would seem, in the case of TV radiation, the only answer would be to remove the television receiver. There are, however, several factors that, with the exercise of good common sense, eliminate any need to take so drastic a step.[38]

It was in the interests of preventing just this drastic step that the manufacturing industry, the mainstream scientific community, and the U.S. government rallied, from 1966 through 1971, to thoroughly downplay the health hazards of low-level radiation and protect this powerful medium from any serious threat to its sacrosanct place in postmodern life. The same forces have also rallied to reassure us about low-level radiation released from underground nuclear tests, nuclear generator "incidents", and, more recently, the Chernobyl disaster.

Since the events of this episode of TV history unfolded, at least two writers have argued that the TV-radiation emission problem has not been solved. In his 1976 Afterword to *Health And Light*, John Ott continues to refer to "radiation from TV picture tubes", while Jerry Mander incorporates this danger as part of his *Four Arguments For The Elimination of Television,* published in 1978.[39] Whatever the case may be, it has certainly been compounded by the widespread use of word-processors and VDTs. (We seem, as a society, to have inculcated some kind of need for a screen in front of our eyes at all times.)

By 1985 the United States had still not established exposure standards specifically for VDT radiation emissions. The technology is not covered by the U.S. Radiation Control for Health And Safety Act of 1968, so VDT manufacturers are not required to submit their product for testing by the U.S. Bureau of Radiological Health. The BRH can only request that manufacturers voluntarily submit their technology for testing, in which case the TV-emissions standard of 0.5 mr / hr. is used as the safety limit.[40] In Canada this x-ray emission limit was applied to VDTs only as of 1979, under the Radiation Emitting Devices Act; the technology, however, had been in use for several years before the Act. Recalling that users of word-processors and VDTs sit just inches from the screen, we have every reason to be fully skeptical about the reassurances issuing from this industry. Since 1946, the forces of government and industry have continually mystified the issue of low-level radiation: first in terms of the bomb-tests, then in terms of "the peaceful atom", television emissions, nuclear reactor safety, VDTs, and most recently, underground nuclear testing and Chernobyl.

As a society, we are meant to patiently endure, and collude with, the new "facts of life". These facts, as the U.S. Atomic Energy Commission's Willard Libby so succinctly reminded us in 1955, increasingly include radiation emissions and fallout: the means for our own mass death. This strange contradiction (the new facts of life equals mass death) becomes the curious paradox implicit in another aspect of the television medium, to which we now turn.

12

The Power of Ghosts

You might think that science and technology have left ghosts in the age of manor houses, but I believe on the contrary that the future belongs to ghosts and that the modern technology of images – cinema, telecommunications, etc. – has increased the power of ghosts tenfold.

JACQUES DERRIDA

BY THE 1960s many feminist, minority, and left-wing political organizations had recognized various aspects of television's patriarchal / white / capitalist / imperialist agenda and were calling for changes in the medium. Usually, the advocated changes focused on overt content (violent programming, sexual and racial stereotypes, ads pitched to children), on ideology (TV's marginalization or exclusion of alternative views), on industry hiring practices, on the American-dominated global distribution of the mass media. In other words, most political activism with regard to television became devoted to gaining access to it so that previously disenfranchised groups and viewpoints – and in the case of countries dominated by "dumped" American shows, their own national images – might reach the vast constituency tuned to the mass medium around the world.

Few critics questioned the very medium itself: its effects as an electronic technology regardless of overt content, its replacement of human vision by the technological cataract, its "language" of visualization (which it shares with cinema) and the ways that visual language reworks both society and our human sensorium. Similarly, few attempted to place the modern technology of images within the larger technological agenda and its "anti-people climate". That deeper interrogation of film and TV is more recent. It raises questions central to the topic at hand: the transformation of all of us into a new species suitable to the technological supremacy of the nuclear age. It is in this sense that viewpoints as diverse as that of Marshall McLuhan, Harvard theologian Harvey Cox, radical filmmaker Peter Watkins, and feminist author Adrienne Rich are beginning to coalesce around that most simple, but radical and difficult act: unplugging the medium.

Early in his career McLuhan recognized that overt programming content was a "red herring", that the real message of TV was its electronic scanning process: a process that he considered deeply violent in its effect on the mind-body. Because of electronic scanning, and because of the TV / brain phenomenon, McLuhan ended his career by urging everyone to "pull the plug" on television. His remarks and explanations, however, were usually too aphoristic to make much sense to people other than those in the TV industry itself, which had early on adopted him as its pop-academic high priest.[1] For his part, theologian Harvey Cox had by the 1980s recognized the "coercive" nature of television and was formulating a view in which "the Christian strategy vis à vis mass media was not to try to use them but to try to dismantle them."[2]

Similarly, by 1979 Adrienne Rich had arrived at a perspective that is worth quoting at length just for its sheer energy in contrast to a TV medium that (in her words) "itself breeds passivity, docility, flickering concentration".

> The television screen has throughout the world replaced, or is fast replacing: oral poetry, old wives' tales; children's story-acting games and verbal lore; lullabies; "playing the sevens"; political argument; the reading of books too difficult for the reader, yet somehow read; tales of "when-I-was-your-age" told by parents and grandparents to children, linking them to their own past; singing in parts, memorization of poetry; the oral transmitting of skills and remedies; reading aloud; recitation; both community and solitude. People grow up who not only don't know how to read, a late-acquired skill among the world's majority; they don't know how to talk, to tell stories, to sing, to listen and remember, to argue, to pierce an opponent's argument, to use metaphor and imagery and inspired exaggeration in speech; people are growing up in the slack flicker of a pale light which lacks the concentrated burn of a candle flame or oil wick or the bulb of a gooseneck desk lamp: a pale,

wavering, oblong shimmer, emitting incessant noise, which is to real knowledge or discourse what the manic or weepy protestations of a drunk are to responsible speech.[3]

Obviously, making TV programming "better" does not rectify all the losses listed by Rich – each of which has to do with not having the left hemisphere of one's neo-cortex effectively amputated by TV while the right hemisphere is colonized, with being involved face-to-face in real life, with not being a spectator.

But even more, as filmmaker Peter Watkins has recognized through research for his anti-media film *The Journey*, the very "language of television or film ... is one structural language conditioning us to accept another set of structures".[4] That other set of structures, which the language of film and TV reproduces in its explosion and re-creation of fragmented time / space parameters via montage, is the machine-world which culminated in the bomb and is organized around the eclipse of the human body and the supremacy of death.

☐

One crucial aspect of the technologically-reproduced image is its "elimination of aura". Simply put, while film and television convey highly realistic images of the world, they cannot capture or convey the aura of living things – that ineffable quality of "presence" or life force that living beings have and dead things don't have. To experience the aura or life force of a living being, you have to physically be in its presence. A substitute for this actual proximity, which film and TV provide through their realistic depictions, cannot convey aura and instead eliminates it. Film and television's elimination of aura has, in fact, been crucial to developments throughout this century and especially in the postwar "new world" steeped in the modern technology of images.

The elimination of aura has long been a minor theme in critical writing on film and photography.[5] Recently, Jerry Mander, author of *Four Arguments For The Elimination Of Television*, has updated that theme by applying it to the television medium.[6] Mander tells the story of working for the Sierra Club in the 1960s when that organization was hoping to establish a Redwood National Park in order to save the giant redwood trees then being logged into extinction. He decided to take a camera into the forest and shoot images of the majestic trees to convey to the TV public the splendour of a redwood forest and just what would be lost if the logging continued. But, says Mander, when the imagery was brought back and viewed, "It didn't work at all. There was absolutely no communication we could have with those trees through technology."[7] The imagery of the live redwoods was flat and unresonant, conveying nothing of the mood of actually being in a redwood forest, nothing of the real splendour and vulnerability of a living thing.

Then Mander decided to go back and shoot images of the acres of redwood stumps that had been left in the path of the loggers. "The images of the stumps were more powerful than the live trees," he says. The audiences that saw those images were shocked and horrified by the destruction of the redwood forest. "That," says Mander, "was when I first realized that death is better for television communication than life is. When you translate images through technology the aura is dropped out and just the image itself remains."

Without aura, the image of a living being is lifeless: merely a detailed husk of an exterior surface. The image may be entirely realistic, "life-like" – indeed, the technology depends on this reproduction of realism – but what it conveys is nothing more than this husk or shell. "That means," says Mander, "that the communication through TV of dead imagery or non-alive imagery is more efficient than the communication of living imagery." Since the dead object (an advertised product, a machine, or a corpse) has no aura in the first place, it loses nothing in the process of being translated into images through technological means. Therefore, "the object communicates 100 per cent efficiently" on the television screen, but "the living thing loses its most important quality, which is its life essence" or aura.

There are programming parallels to Mander's Sierra Club example. Canadian television has a number of nature shows on the air, and has always favoured this content. The paradox of TV shows such as *The Nature of Things* or *Wonderworks* is that they attempt to convey the wonder of living nature on a medium that eliminates aura. Thus, the most successful of *Nature* host David Suzuki's programs are arguably those which, like Mander's images of the redwood stumps, show the destruction of living things. For example, the last show of the 1986 season of CBC's *The Nature of Things* focused on acid rain. The images of the dying forests in Germany spoke powerfully in a way that the images of Canada's threatened, but visibly undamaged, forests did not. The dead stumps and dying forests communicate, in Mander's terms, with 100 per cent efficiency on the screen, while the images of an undamaged forest or lake are unmoving beyond the sentimental because the technology cannot convey their greatest characteristic: life force or aura.[8]

TV's elimination of aura is perfectly suited to the medium's role in the postwar world, most immediately in terms of the morality of consumer hedonism advocated by Ernest Dichter and the new science of motivation research. Imaged on screen, a living being loses its greatest characteristic and is thus reduced in stature by the process. But an object, which never had an aura in the first place and so has nothing to lose, can be given a surrogate "life", an artificial "aura" – as advertisers do for their products when they have them move and perform on screen, or even sing and dance and appear to be "living" things. Mander writes:

> These factors conspire to make television an inherently more efficient and effective medium for advertising than for conveying any information in which life-force exists.... So television accomplishes something that in real life would be impossible: making products more "alive" than people.[9]

Postwar consumer hedonism, the manufactured desire for increasingly more objects, was founded on the image, not on the body. The body has its natural satiation points, its limits to consumption, its sense of enough. But in a society based on the image, there are no limits to which consumer hedonism may be pushed. Since the TV image makes products more "alive" than people, it is the perfect medium for not just a product-centred ethos, but an object-centred culture. As a mass medium of spectacle located in the home, television is ideally suited to (and was intentionally launched in) an age in which objects and technologies have more value and worth than living beings.

Since TV best conveys the aura-less object, conferring upon it a surrogate "life", it is not surprising that much of TV's overt content is a celebration of the machine world. Images of technological devices dominate programs, whether in speeding car chases, the wonders of spacecraft, robotics, computers, gadgets, military hardware, and TV technology itself. Shows geared to children are especially populated by machines, and a program such as *Knight Rider* features a technological object (the talking supercar) as the sidekick for the hero, replacing the human buddy. TV news is especially a news of machines, with an obvious visual emphasis on their actions and movements: planes landing and taking off, firetrucks arriving at the scene, warships manoeuvring at sea, helicopters bringing politicians to meetings, rockets blasting off, weaponry on display, and various technologies at work in the background of many on-screen interviews.

Perhaps the most stunning recent example of the medium's technological bias has been the reporting on Reagan's proposed "Strategic Defense Initiative". Inevitably, any television news reference to "Star Wars" is accompanied by hyper-realistic animation of rays zapping incoming missiles, videogame-like explosions eliminating enemy blips: seductive imagery that fascinates, makes the prospect appear tidy and efficient, and also presents the technology as a virtual fait accompli. As Todd Gitlin has written, such imagery, with its "cheerful visual language", fosters the assumption "that what can be drawn can be planned, what can be planned can be

built, and what can be built can protect. The diagrams helped confer upon
SDI the force of the feasible; the opposition looked like progress-bashing
grumps."[10]

But we must not lose sight of the fact that the medium itself, as a technology that eliminates aura, also fosters the centrality of other technologies, other objects. Its entire bias, even when it depicts living beings, is the
dead object.

Since the technologically-visualized human being is devoid of life force
or aura, it communicates less well than the object unless it is given an
artificial "aura". That is the function of the publicity machine in our society. As long ago as the 1930s, Walter Benjamin noted this in his discussion
of filmmaking and the elimination of aura. As he wrote, Hollywood
depends on "an artificial build-up of the 'personality' outside the studio".
According to Benjamin, "The cult of the movie star, fostered by the money
of the film industry, preserves not the unique aura of the person but the
'spell of personality', the phony spell of a commodity."[11] Much of print
journalism, with its emphasis on media celebrity, serves to enhance this
spell, while guest-appearances on TV talk-shows and other programs
increase the commodity status of each "personality". The non-celebrity
achieves momentary commodity status by being imaged on the TV screen,
especially if she or he is in proximity to someone (a game-show host, a
reporter, a talk-show host) whose familiar image signifies TV itself. In the
same way, the news anchorperson takes on the surrogate "aura" of TV
technology and becomes the signifier of live transmission.

More importantly, as Mander's comments indicate, because death and
dead objects communicate better on screen than life and live beings, the
medium *as medium* has a necrophilic bias. Program-makers have (consciously or not) utilized this fact in their emphasis on the spectacle of the
carefully orchestrated car crashes, explosions, and full-scale disasters that
characterize so much of TV entertainment. But this is not to suggest that TV
program content should be otherwise. It is, rather, to point out that death,
and surrogate "life" for dead objects, go with the territory.

□

This aspect of the elimination of aura is central to understanding the place
of the modern technology of images within the larger anti-human, anti-
body agenda. We may explore it further by turning to an essay by film
critic André Bazin. Likening photography to the death mask, Bazin notes,
"Photography does not create eternity, as art does, it embalms time, rescu-
ing it simply from its proper corruption."[12]

This embalming of time in the photograph removes the image from the flow of life and inserts it into a different realm. Contrasting photography with painting, Bazin argues that the two processes are entirely different: while both yield an image, the painterly process fully involves a human subject working through a medium such as oil to express a vision. With photography, Bazin suggests, "For the first time, between the originating object and its reproduction there intervenes only the instrumentality of a nonliving agent", the camera.[13] While one could argue with this contention (and others have), for Bazin this central point has to do with both photography's elimination of aura and its automation as a process.

Standing in the presence of an original painting by Van Gogh or Georgia O'Keefe, for instance, a viewer senses that the aura of the painter has somehow been inscribed onto the canvas. This has nothing to do with its hallowed setting within a gallery or museum. Rather, it is a quality, a characteristic of the original painting as in some sense a living thing. With photography, however, there is first of all no such thing as "the original work", unless it is the negative – the work of light interacting with silver haloids. In the resulting photograph, we perhaps detect the photographer's style, but not her or his aura, as we can in the presence of an original painting. Moreover, a human photographer need not even be present to get a fully realistic photograph. Think, for example, of photomats where the insertion of a coin is enough to produce an image. As Bazin writes, "All the arts are based on the presence of man, only photography derives an advantage from his absence."[14]

This absence is central to our discussion of the nuclear age. It is an absence that occurs both at the site of the image (the aura of the living being is eliminated) and at the site of the originating of the image (no human agent is necessary, or is necessary only as a servo-mechanism in the process: tripping the shutter). Modernity is characterized by this absence of the human. Its "vision" is one in which humanity and living beings are unnecessary and absent, except as traces documenting technological progress. We might say of the fully technological society what Bazin said of photography. It is the only society that derives an advantage from humanity's absence.

Historically, the invention of photography (and then cinematography) in the middle and late nineteenth century occurred at a timely moment in the advance of patriarchal capitalism. In its ruthless obsession with power and wealth, the patriarchy had reached the point where its purely linear, emotionless, goal-directed efficiency could begin to conquer time and space through the mechanical dynamo: an efficiency and a conquering dependent on a particular definition of "progress" that would obscure the interests of capital, make a fetish of technology, and "naturalize" the subservience of the body to the machine world.

Patriarchy's forceful linearity and dreams of "progress" cut a swath across continents and consciousness itself. In its wake, the slower, natural ways of being and doing – as well as entire indigenous populations – toppled like forests. Not surprisingly, photography and then cinema emerged to cast a mechanical "objective" gaze upon this march of progress. As Marx stated, "It would be possible to write a whole history of the inventions made since 1830 for the sole purpose of providing capital with weapons against working-class revolt."[15] Arguably, one of the best weapons was the photograph, with its elimination of aura, its fascinating realism, its advantage derived from the absence of man. As a death-mask embalming time, its "vision" announced the end of space / time based on the human body and nature.

During the 1870s, an English photographer, Eadweard Muybridge, began experimenting with the possibility of making a sequence of still photographs that, when projected at rapid speed, would realistically imitate the movement of living beings. This desire in itself indicates the fascination with simulacra, corresponding to patriarchy's inherent loathing of the actual body, and capitalism's technological fetish. British film scholar Steve Neale quotes from a press report of November 29, 1881, which lauded Muybridge's work:

> Imagine: with the telephone (read "phonograph") you can already preserve the human voice in a box, just like sweets; with a series of animated photographs you will be able, years later, to rediscover the way a man moves or holds his head. The ghost will walk, and that is how little by little, science, progressing with giant steps, will succeed in abolishing death, its sole obstacle and only enemy.[16]

The last sentence of this 1881 report indicates that life had already been replaced by the image, whose moving pictures would "abolish death". Named here as the "sole obstacle and only enemy" of science, death haunts nineteenth-century invention as fully as it haunted Shelley's Frankenstein, perhaps an archetypal prefigurement of Muybridge as scientist-engineer. Just as Frankenstein's desire to abolish death was itself tinged with death and constellated it around him, so, too, this 1881 report speaks of preserving the human voice in a box, as a corpse is preserved through embalming. And it holds out the tantalizing possibility that "the ghost will walk". What is difficult to see is why this 1881 writer would consider the image-"ghost" walking on screen to be a sign that death had been abolished: unless we see the moving images' realism as having eclipsed the reality of the living body. A later press quotation from 1895 amplifies this:

> When at last colour photography has been achieved, and when a phonograph has been added to that, then movement and speech will be caught

simultaneously with rigorous accuracy, that is to say life itself. When that day has come – and it is coming tomorrow – science will have given us the complete illusion of life. Why shouldn't it be capable of giving us life itself?[17]

The quotation indicates the developing ideology: "rigorous accuracy" is equated with "life itself". While it also admits that these simulacra provide "the complete illusion of life", the 1895 press report considers that illusion as presaging science's ability to provide "life itself". We must read this last rhetorical question from two angles.

In his excellent text, *Fathering The Unthinkable: Masculinity, Scientists And The Nuclear Arms Race*, Brian Easlea traces in the language of scientists the desire to imitate the life-giving capabilities of the female. He particularly traces this desire in the extensive birth metaphors surrounding the creation of the atomic bomb, especially in the coded language used to signal the success of the Manhattan Project in its first testing of the bomb in New Mexico and before its delivery to Japan. This "simultaneous appropriation and denial of the feminine" has long been apparent in scientific goals.[18] The irony of that appropriation in the creation of the bomb, the epitome of mass death, serves to indicate the unconscious hostility to the feminine and to life that underlies modern science. But we can see this appropriation of what is uniquely female – the ability to give birth – in that 1895 press report heralding the possibilities of cinema, and through such simulacra of "rigorous accuracy", the assumption that science must be "capable of giving us life itself". The realistic moving image was thus the sign of patriarchal capitalism's potential to eliminate the female from the birth process. If science could provide the "complete illusion of life", why not life itself?

But we might also read that 1895 rhetorical question from another angle: as an admission that life has been made absent, replaced by illusion. If life and the human body were fully present, there would be no need to look to science to "give us life itself", to fill this lack. We must equate this absence not only with the modes and methods of capitalist industry, but also with that absence that Bazin alerts us to: the absence of the human being and aura from photographic reproduction.

Already working at the pace of the factory machine, which has nothing whatsoever to do with bodily rhythm and needs, the population of the late nineteenth century was undergoing a further fragmentation and mechanization of the body through the emerging spectacle of cinema. This new simulacra fully exploded the human dimensions of time / space through montage. It replaced human vision with the disembodied camera-eye: we identify with this machine-gaze as we watch the screen. Freed from the temporal and spatial constraints of bodily vision – which depends on the actual movement of the whole body in relation to the object seen – the cinematic eye constructs a disembodied time / space, a technological "vision" that fascinates precisely because of this conquering of bodily limits. Already robbed of life and the human body by the workings of capitalist patriarchy, under which we function as extensions of machines and as commodities bought and sold for our labour, we are further objectified and robbed of our actual bodies by technologies of sensory reproduction – which mechanically reproduce not just "vision" and the voice in the box, but us as spectators and reprocessed beings: with our senses effectively auto-amputated from the body and reorganized on a technological plane that suits a larger agenda.

□

The twentieth century has been characterized by the increasing technological perfection of "the complete illusion of life". But this patriarchal desire has been accompanied (perhaps necessarily) by the utter deterioration of life itself: the natural environment, meaningful social ritual, satisfying and non-alienating ways of living and working, even the quality of those fundamentals that human bodies need just to subsist – air, water, food. The irony of this simultaneous perfection of realism / deterioration of reality is itself a dislocating violence perpetrated on the human species: a violence that must be traced to that central loathing of the body that characterizes patriarchy.

Marshall McLuhan was not interested in questions of patriarchy, capitalism, or U.S. imperialism, but he did attempt to explore the technological colonization of the human body by electronic media. As he said towards the end of his career, "Electronic man is in desperate need of roots. He's lost his body, his physical body and his private identity. He has an image, but no body."[19] He warned of the violence that springs from this loss of the body – that people kill to find out if they're actually alive, seek real death to replace a living death.

But it is precisely as "nobodies" that we are intended to exist under the sign of the mushroom cloud. The mass of humanity has become utterly dispensable, no more necessary to the expanding technological systems than the human guinea pigs vaporized in Japan or marched onto the Nevada desert for the bomb-tests. Fittingly, the postwar electronic age pulsates not to bodily rhythms but to the rapid-fire stream of electrons directed by cathode-ray guns in the TV picture-tube: aimed through the viewing screen at us. As feminists Monica Sjöö and Barbara Mor have observed, "The way to control human life is to control the rhythm of life".[20] Even more effective is to replace it with a fully technological rhythm approaching the speed of light. That triumvirate of techno-systems that rule the postwar world – television, nuclear weaponry, the computer – is a triumvirate dependent on electronic circuitry: advances in which have now combined all three into a missile guided by computer and providing visual feedback to the pilot via built-in video camera. In the age of electronic spectacle, war has become a video-game.

Roland Barthes has written:

> Instead of constantly locating the advent of Photography in its social and economic context, we should also inquire as to the anthropological place of Death and of the new image. For Death must be somewhere in a society: if it is no longer (or less intensely) in religion, it must be somewhere else; perhaps in this image which produces Death while trying to preserve life.[21]

If death is in the image, then with television that death-image is more fully consumed by us than is the case with either the still photograph or the projected film. The scanning process depends on the viewer's completion of the mosaic pattern (at electronic speeds). Our brain fills in the complete image-gestalt, which is otherwise never entirely present on the screen. Television thus takes to another level the absence discussed by Bazin. We consume the video-void, even as we are consumed and processed by it. It was in this sense that McLuhan perceived television to be the equivalent of the bomb. In his aphoristic fashion, he wrote of the postwar electronic age: "The Bomb itself became content, having had a short reign as environment".[22]

Perhaps not surprisingly, the one visual technique unique to television – the instant replay – arose out of a death. Human vision, and even film technology, has no such ability to embalm an instant and replay it instantaneously, without the need of chemical processing (that is, a lapse of time, as with film). The live TV transmission of the removal of Lee Harvey Oswald from his Dallas cell in 1963 meant that TV cameras captured on (the then-new) videotape technology his murder by Jack Ruby. The moment was instantly replayed for the viewing audience – death being the greatest moment of spectacle. The industry subsequently adopted the technique for other highlights.[23]

Writing in the 1930s, Walter Benjamin detected the profound changes being worked on the human organism by the mass technology of film and warned that this fragmenting, restructuring, and replacement of the bodily sensorium could only "culminate in one thing: war".[24] Of the mass audience he wrote: "Its self-alienation has reached such a degree that it can experience its own destruction as an aesthetic pleasure of the first order."[26] But Benjamin (who may here have predicted World War II) was writing before the invention of two new global mass media: TV and the atomic bomb. Perhaps we must now update his observation to refer to planetary annihilation.

From this perspective, ABC's *The Day After* (broadcast in November 1983) achieved a notable level of television perfection. Since the unwritten and unspoken *gemeinschaft* of program production is that TV shows should not move audiences emotionally in any deep way (since that would interfere with the commercials), the broadcast of *The Day After* (duly sponsored) indicates that as a mass TV audience we have reached that point predicted by Benjamin. ABC's publicity slogan for the broadcast seemed to incorporate the recurring paradox of the new world: the "facts of life" dictate that we become accustomed and adjusted to mass death through radiation. "*The Day After* – Beyond Imagining", hyped the ABC publicity machine, which then continued incongruously: "The starkly realistic drama of nuclear confrontation and its devastating effect on a group of average American citizens will air November 20th."[26] This event "beyond imagining" provides the images, "starkly realistic" of course, of mass death on a medium that has long been part of the pro-nuclear campaign itself.

"Although based on scientific fact the following film is fiction," stated a caption introducing the prime-time transmission. "Because of the graphic depiction of the effects of a nuclear war, viewers may find some scenes disturbing." As Walter Benjamin predicted, in the age of alienation from the bodily sensorium, the mass audience can experience the spectacle of its own destruction as, if not "an aesthetic pleasure of the first order", then at least as a pleasurably disturbing diversion, and one in which, thankfully, the unleashing of the bomb annihilated all commercial messages. ABC saw fit to schedule no ads during the post-attack portion of the broadcast.

Jacques Maritain has named our mass media era "the Age of Angelism", an appropriate metaphor to connote the loss of the body.[27] Jacques Derrida speaks of "the power of ghosts" in an age dominated by the modern technology of images.[28] Both metaphors also refer to the necrophilia inherent in the technologies, which were ironically intended to "preserve life" and "defeat death". As with Frankenstein, the intention constellates its opposite. In the death-images of film and TV, conjoined by editing and montage into a re-creation of time / space, we may find the equivalent of Frankenstein's undead monster, sewn together out of pieces of corpses, animated by some superior force harnessed by the scientist-engineer.

But there is a certain naïveté in Shelley's novel, or rather, it is simply a product of its time (1818). Mary Shelley, after all, was writing well before the splitting of the atom: which may have been nothing less than a shattering of the world. That the onset of research and effort to split the atom coincided historically with the invention of cinema in the late 1880s is a fact that should alert us to the fundamental change that occurred simultaneously with the great advance of capitalism. All relations were being radically rearranged into a new order that culminated in the atomic bomb.

Within that "new world", Mary Shelley's scientist-engineer portrait seems too repentant, too emotional, too human for the global efficiency instituted under the sign of the mushroom cloud. The twentieth-century scientist would have to be one who could prefer the "technically sweet" over the imperfect human body, who could see technology as the new evolutionary principle. Unlike Frankenstein, the scientist would not be appalled by technological simulacra, but would instead see them as far more wondrous than anything nature had wrought. This new scientist-engineer would especially be able to look upon mass death with the cool eye of objectivity: calculating the degrees of human expendability according to the larger, more efficient goals of the expanding technological dynamo, whose priorities are maintained through the image.

Those priorities have long been clear, but we might focus on a particular event of 1951 to crystallize them. On January 29, 1951, radioactive snow blanketed the film production plant of the Eastman Kodak Company in Rochester, New York. The high radioactivity threatened to ruin the plant's film-stock. The company traced the fallout to a Nevada bomb-test over two thousand miles away and lodged an immediate complaint with their industry watchdog, the National Association of Photographic Manufacturers, as well as with the Atomic Energy Commission.

> The Kodak Company's first inclination was to sue the government for damages. They settled, instead, for privileged status: in an extraordinary departure from a policy of secrecy, the AEC offered to send Kodak, in advance of all new tests, a set of secret fallout maps predicting the areas of possible heavy radioactivity. Several top executives of Kodak and other photographic companies were given security clearances so that they could handle the information. Those who lived in Nevada and southern Utah, and who were about to be more directly affected by the fallout than the Kodak Company at Rochester, were not so privileged.[29]

In an age structured on the twinned necessities of mass spectacle and mass annihilation, an age dominated by the image as a locus of death, the Kodak-AEC settlement contains a certain logic.

13

The Astronautical Body

THE TWO-MINUTE TV clip opens with a long shot of the spacecraft leaving the launchpad and climbing steadily up into the sky, its booster rockets billowing massive trails of white smoke behind it. On the audiotrack Mission Control and Shuttle Commander Scobee exchange data about APU's (auxilliary power units), velocity, altitude. Cut to a telephoto close-up on the right side of the space shuttle Challenger, and the words from Mission Control, "Challenger, go with throttle up."

"Roger, go with throttle up," confirms Scobee. Cut to a wider view of the shuttle, which suddenly bursts into a huge fiery gold and white cloud with two strange Y-shaped tendrils shooting off and down across the blue sky. Forty seconds of silence follow as the TV camera pans with the exploding debris.

On Tuesday, January 28, 1986, this two-minute TV clip dominated the airwaves of North America. Although the only network covering the Challenger launch live was Ted Turner's Atlanta-based Cable News Network, within six minutes of the disaster CBS, ABC, and NBC, followed by CTV and CBC, broke into their regular programming and stayed with live coverage for five hours straight: playing and replaying again and again this eerie two-minute video clip.

Time magazine called it "a nightmarish image destined to linger in the nation's shared consciousness". Senior writer Lance Morrow stated: "Over and over, the bright extinction played on the television screen, almost ghoulishly repeated until it had sunk into collective memory. And there it will abide, abetted by the weird metaphysics of videotape, which permits the endless repetition of a brute finality." CBC's *The Journal* called this two-minute video clip an "apocalyptic image". Writing for *The Toronto Star*, Joe Erdelyi referred to the need to "wake up to the reality of what the screen portrays with such cold artistic beauty".

It is perhaps this last point, the "cold artistic beauty" of the TV imagery, which most deserves comment, and contemplation.

Unlike many people, the only TV coverage of the Shuttle disaster that I watched was provided by CBC's *The National / The Journal* at 10:00 p.m. the evening of the event. But even there, the two-minute video clip was replayed at least four times. What immediately struck me while watching it was the strange, uncanny, aesthetic beauty of the TV imagery: like some perfectly shot sequence of fireworks in summer. Maybe because of the smooth, even panning of the TV camera as it followed the exploding parts across the sky. Maybe because of the forty seconds of silence beneath the sequence: highlighting the spectacular quality of the visuals. Maybe because I've watched too much TV.

All I know is that, as Joe Erdelyi observed, this two-minute clip did convey to me a "cold artistic beauty" devoid of any emotion save for a kind of technological awe. Not an awe of technology, but rather a machine-like awe towards the performance – even the spectacular failure – of another machine, and the success of the camera-eye witnessing it.

It was only when I heard another human voice actually expressing the cold, emotionless void I had momentarily fallen into, that I was able to snap out of a technological fascination with the imagery. Steve Nesbitt, the commentator at the Johnson Space Center in Houston, had paused for those forty seconds of silence while the television screen filled with the awesome exploding pyrotechnics. When he resumed his public narration, it was to say, in a voice completely devoid of emotion: "Flight controllers are looking very carefully at the situation. Obviously, a major malfunction."

The shock of Nesbitt's utterly technological response snapped me back into the land of the living and the dying. Up until that point, the seven crew members had momentarily faded from my consciousness: completely overshadowed by the cold, awesome, artistic beauty of the TV image.

I suspect that my response is not atypical, that many people watching their TV screens experienced this void of technological fascination, this draining of their own human response in relation to the TV spectacle of awesome exploding disaster. In this sense, CBC's *The Journal* was right: it was and is an apocalyptic image.

While others claim that the visuals of this two-minute clip will remain

forever embedded in their minds, for me it will be the efficient, emotionless, technological voice of Mission Control calmly stating: "Obviously, a
major malfunction."

□

Aside from the gargantuan nuclear industry, the other alluring priority of
the patriarchy (in all its present ideological forms) is the space industry.
Digging beneath the technological jargon and the patriotic rhetoric that
surrounds this industry, one finds stark indications of the philosophy that
fuels the spending of $8 billion a year on space technologies: a philosophy
that is quite clearly in line with the patriarchal loathing for earthly nature.

On April 12, 1961, Yuri Alekseyevich Gagarin made that first historic
orbit: a leap (in the words of a newspaper text) "beyond the surly bonds of
earth". Over twenty-five years later, reading of this past exploit, I cannot
help but think of the fate of the seven Challenger astronauts in a television
event witnessed by millions. Nevertheless, the collective agony did not
generate (judging by the media, at least) a single voice raised in question of
the space program itself. I gather that it has now become utterly taboo to
express doubts about the sanity of the whole enterprise. Rather, the
assumption, which my newspaper dutifully reiterates, is that "the human
race has been emerging like a butterfly, struggling to tear itself free from
the cocoon of its planet".[1]

In this rhetoric, the reporter is obviously echoing the voices of the space
industry itself. "The move of humans off the face of the planet," says
Robert Jastrow, former director of NASA's Goddard Institute for Space
Studies, "is the most important step in the evolution of life since the fishes
left the water 350 million years ago. Science is not the space program's
only purpose – or even its main purpose."[2]

"Living on the moon will be a wonderful, invigorating experience,"
enthuses Neil Ruzic, a space-applications consultant. "There will be
people who will stay, and their children will call the moon home. Earth
will be an alien world they may never want to visit."[3]

"I believe that space," states Peter E. Glaser, space-business consultant,
"in the 21st century will probably be what aviation, electronics and computers were, together, in this century. It is the next evolutionary step for
humanity."[4]

The recent IMAX film, *The Dream Is Alive*, released in advance of the
Shuttle disaster, is a thirty-minute ad (disguised as a documentary) for
NASA. Through the advanced IMAX technology, the film (like all IMAX
films) must always be shown in special cinemas built with a viewing screen
six storeys high.[5] Utterly dwarfing the audience, the images are so huge,
and the sound so all-encompassing, that one feels completely eclipsed by

this display of hyper-realism on the gargantuan screen. In the case of *The Dream Is Alive*, this extraordinary technology of spectacle seems perfectly suited to the goal of hyping the space industry, which – as institution and ideology – is similarly structured around that Bazinian "absence" of the human. The film's narrator is no less a signifier of authority than Walter Cronkite, who waxes optimistic on the prospects of our grandchildren being born in space, while celestial voices on the sound-track underscore, in full religious chorus, the Christian righteousness of this project.

The fervour and media hype are entirely typical of the thinking within the space fraternity. Since the 1950s this fraternity has fixed its gaze on achieving what it considers to be nothing less than the next evolutionary step for the species, or at least for a technologically elite part of it. The statements emitted reveal, to my way of thinking, a mind-set that perceives the earth itself as passé: an old, worn-out environment to be left behind, discarded like a useless husk or empty packaging. For such a mind-set, the fate of the earth must, in the long run, mean next to nothing. The intention is to abandon it anyway and go off to colonize space.

The space industry is, of course, the flip-side of the nuclear weapons industry. It is the escape hatch through which a select few will survive a doomed planet. Indeed, by holding out to us the spectacular possibility of human life transported elsewhere, space-industry hype tends to replace passionate concern for the fate of the earth with an excited urgency to get those space stations and transporter vehicles ready in time. I sense that on some deep level, the continuous threat of nuclear war since 1946 has fed, and still feeds, the patriarchal impulse to fully abandon the planet: carrying out the war against earthly nature to its next logical step and its next frontier.

Like the nuclear industry, the space industry has also depended on the medium of television to bring its own form of spectacle into our homes and thereby foster an enthusiasm for every space endeavour. In the celebrated figure of the astronaut, always the focus of media attention, we find the emblem of patriarchal enterprise and its centuries-long dream of eliminating the loathsome human body, replacing it, like Frankenstein, with something more perfect.

The figure of the astronaut fully suited up reveals a new, gleaming white "body" appropriately tattooed with the NASA logo and ready to risk itself on the pyres of progress. The human mind is here enveloped in a new casing: not "strange and repugnant" like its earthly predecessor, but smooth, plump, impermeable and masked – a better (synthetic) skin for this mind so superior to brute nature. While the other sensory functions are suitably mediated or swathed beyond input, the face-mask allows that most privileged of all human senses in the twentieth century – vision – to dictate to this fully digital creature, who has been thoroughly trained to override all emotion with a logical practicality geared to complete efficiency.

This new body, surrounded by gleaming, metallic technologies that mirror its synthetic skin, glides weightless beyond the earth. Its age-old tie to the natural order, to the planet itself, has been severed and replaced by a plastic umbilical cord which connects it to its new TechnoMother: the computer that monitors all its functions and flies the astronautical body in a synthetic womb out beyond the pull of the earth.

This astronautical body, in its perfect weightlessness, its manufactured skin, and its trusting dependence upon machines, has at last overcome that clumsy impediment to progress, that embarrassment to patriarchal efficiency, that thing of madness and scandal which was the earthly body. This new body is quite literally a no-body: a series of blips on the computer screen, an image on the TV monitor, a servo-mechanism functioning in perfect sync with the machines around it. And, having been transformed into this obedient servo-mechanism, the astronautical no-body thereby achieves the status of subject: worthy of investment, protection, and the media hype suitable to all space and military hardware.

In the figure of the astronaut, we find the new hero of the nuclear age: the completely masked man, rootless and floating, impermeable and disconnected (except for technological mediations), fully beyond circadian rhythms and influences, ungrounded, efficient, and completely monitored and dependent on technology.

This astronautical figure, speeding on a trajectory into the void, looks back upon the earth and, seemingly for the first time, sees that from this extraordinary distance, from this height of technological achievement, the earth is whole and even lovely. For the patriarchal mind, it is an insight that can perhaps only be gained from this god-like perspective, this hovering beyond the surly bonds of the planet. Before turning back to the future, the astronaut aligns the small blue-green planet within the cross-hairs of the video-camera: recording, like a tourist, another crumbling monument.

14

Conclusion:
The Whirlwind, The Storm

In the long run the greatest hope of mental health in the future of the
peaceful uses of atomic energy is the raising of a new generation
which has learnt to live on terms with ignorance and uncertainty,
and which, in the words of Joseph Addison, the 18th century English
poet, "rides in The Whirlwind and directs The Storm".

THE WORLD HEALTH ORGANIZATION, 1957

THE UNITED NATIONS' WHO prescription for "mental health" in the nuclear
age – a prescription advocating psychological denial and ignorance –
seems to have been followed to the letter by the television industry in
North America. We have especially been kept ignorant of the nuclear arms
industry, its relation to atomic power plants, and the dangers of radiation
from a wide range of man-made sources: an ignorance maintained by television for the benefit of the military-industrial institutions. Since TV provides the ideological frame for such institutions, and was itself an out-
growth of the nuclear industry (in its mutual corporate backers and its
own radiating properties, at least during the medium's first two and a half
decades), television has been the perfect machine for constructing a mass
populace united around the bomb.

Ever since 1945, the P.R. campaign for U.S. nuclear weapons expansion
has depended on a radical split in the representation of the atomic bomb:
an ideological separation of two strands of imagery / information. The

162

official representation of the bomb has focused on technological spectacle: the bomb as the culmination of scientific achievement, the harnessing of power by democracy for democratic ends. This representation of the bomb as supernaturalized abstraction of American ideological power has ruled out that other strand of imagery / information that reveals the bomb as the culmination of designed affliction of suffering on the human body at a mass scale. The purposeful ideological separation of the bomb from the body is what has allowed the nuclear industry to proceed unabated for forty years. It is quite probable that if the American public had viewed the film footage of the irradiated Japanese survivors of Hiroshima and Nagasaki, that public would have had a different view of both the "necessity" of atomic weaponry and "the sunny side of the atom".

Nevertheless, that ideological split in the bomb's representation continues to this day. Patriarchal capitalism must continue to hide the bomb's effects on the body in order to maintain the ignorance and psychological denial useful to its purposes. As it has throughout its history, television maintains this ideological split at the level of overt content. The split is graphically illustrated by the different fates of two film productions in the 1980s: *If You Love This Planet* (Canada) and *The Day After* (U.S.).

□

In 1982, Terri Nash, a Canadian film director for the National Film Board's Studio D, made a film called *If You Love This Planet*. Using footage of a speech by Dr. Helen Caldicott (U.S. National President of Physicians for Social Responsibility), as well as some of the long-censored footage of Japan's radiation victims, Nash's twenty-six minute documentary makes vividly clear what the bodily and environmental effects of detonating a single twenty-megaton bomb would be. The images of the Japanese survivors of the bomb are appalling and unforgettable: human beings who, seven months after the bombings, have massive radiation burns, gaping wounds where the flesh has been seared from the body, terrible diseases and deformities that pay witness to the effects of radiation from a single bomb (and a small one compared to today's weaponry). This is the visual material repressed and censored by the U.S. military until 1980.

The work of Jerry Mander has alerted us to the television medium's inherent bias towards death and destruction. Given this bias, one might think that Nash's film would be a natural choice as TV content. But the fate of the film reveals much about the *particular kind* of death-image that our techno-institutions prefer. CBC-TV repeatedly refused to show *If You Love This Planet,* claiming that the film was "biased". It wasn't until Nash's documentary won an Academy Award that the national network finally agreed to broadcast it, unannounced and with no advance publicity, on the night of the Awards. Thus, the broadcast, virtually surreptitious, could escape the advance ire of the pro-nuclear lobby, and the film could be safely contained within a context of Hollywood hoopla and celebrity. This single showing (over five years' time) on the English-language network thus reached a far smaller audience than would have been the case if CBC had publicized the broadcast in advance.[1]

For its part, the U.S. Justice Department has maintained, from 1982 up through 1987, that the film is "political propaganda" under the Foreign Agents Registration Act. Any U.S. screenings of the film must open with a disclaimer warning of the lack of U.S. government approval, and a list of all viewers' names and addresses must be sent to the U.S. Justice Department. Obviously, such restrictions have very effectively prevented the film from being broadcast on U.S. TV screens.

Although television as a medium is biased towards death and destruction, when it comes to the televised depiction of the bomb there is a particular kind of image of death and destruction that is allowed. One year after Nash's film was released but prevented from reaching the mass TV audience, ABC broadcast what must be seen as the spectacular substitute for her documentary images – The Day After. Once again, as has been the case with U.S. imagery of the bomb since 1945, the focus was on the sheer blast-power of the weapon – in this case, within the four-minutes-long strike on Kansas. Comprised of actuality footage of firestorms and atomic blasts, as well as special-effects sequences, this highlight of the program was arguably not horrific but rather, aesthetic: a visual spectacle of destruction carefully designed for the screen. A number of the program's critics noted the contradiction between the "beauty" of the images and the knowledge of the destruction they represent – a contradiction central to spectacle itself.[2] By contrast, If You Love This Planet reaches right to the bodily core of each viewer and has no need for special effects. The bomb itself has created its special effects on the bodies of those who stood before the 1946 cinematographer and wordlessly revealed their appalling wounds.[3]

Not surprisingly, The Day After conveyed the idea that a nuclear holocaust is survivable. Days after the strike, its fictional U.S. President addressed the nation by radio broadcast, saying: "America has survived. There has been no surrender. We remain undaunted." The TV production thus encapsulates that peculiar American paranoia and victim-stance that has been the underlying theme of the U.S. media ever since the bombings of Japan.

In advance of the much-publicized broadcast, the makers of The Day After stressed their commitment to waking up the American public to the dangers of the arms race. But if the television industry were actually committed to an anti-nuclear stance, it could do something as simple as show-ing the footage of the irradiated Japanese. No other images would be more powerful. There is no need for a fictionalized representation of an event "beyond imagining". The actual images of the bomb's effect have been in existence for forty years. Terri Nash used just a few of those images in her documentary, with the result that virtually the entire battery of institution-alized power (including the U.S. Supreme Court) rose up to ensure that the mass public would not see those images. Meanwhile, ABC's TV spectacle reached one hundred million American viewers and countless millions around the globe – becoming the most watched made-for-TV movie in history.

The differing fates of these two productions reveal an underlying code for televised imagery of the bomb in North America. *If You Love This Planet* transgresses that code, while *The Day After* does not. The code is the same one that has been in effect since 1945 with regard to the official representation of the bomb. Imagery of atomic blast-power that appears unrelated to the body is acceptable. *The Day After,* focusing on that blast-power's spectacular destruction of buildings and objects, rather neatly avoided radiation's effects on the body, implying that the worst might be the loss of hair. But more importantly, the production as fiction centred around the depiction of the U.S. as victim. In both these aspects, *If You Love This Planet* is a clear transgression of the code.

The images in Nash's film are undeniably historical, non-fictional in origin. The sequences focusing on the irradiated Japanese refer to a specific historical past, not a fictional future. Moreover, those sequences remind us of a past in which North America was clearly the nuclear aggressor, not the victim. As we gaze at the effects of our bomb on the "other", it is not pos-sible to feel any sense of victimized paranoia. Instead, one feels a sense of guilt for what was done, and a refusal to let it be done to any other popula-tion. This distinction is a key one, and crucial to the dominant code around the bomb. *The Day After* focuses on what could be done to us. The result-ing paranoia is quite useful to justify building more weapons and bombs for our "protection". *If You Love This Planet* shows us what we did to them.

The second transgression of Nash's documentary is that, in the age of the loss of the body, *If You Love This Planet* makes a simple appeal for what might be called sympathy for the body. Unless one has been com-pletely destroyed by patriarchal capitalism's cynical exploitation of the body by the image, it seems impossible to view the images of the irradiated Japanese and not feel a bodily compassion, a solidarity at the level of the flesh. By reconnecting the bomb and the body in the historical images from the past, Nash's film undercuts the technological spectacle of the bomb, that sense of it as an abstraction of power and scientific achievement. Instead, it reveals the bomb to be the epitome of patriarchy's hatred of the body. No wonder her film must be kept off the TV screens of the continent.

□

Patriarchal capitalism depends on the loss of the body as the basic societal condition. Unlimited technological innovation proceeds on the basis of machine values and technological "needs", rather than human values and bodily needs. Moreover, the loss of the body is a condition that is endlessly exploitable, since it generates an unnamed despair, a panic, a hunger that can be labelled as any number of substitute "needs" and "desires" – including the "need" for more nuclear weapons.

According to a growing number of psychologists and feminists, the trust, love of, and grounding in one's own body originally must come from a maternal matrix, a mother, who is herself firmly grounded in bodily wisdom. But in the course of the four thousand years-long development of the patriarchy, the mind / body, spirit / matter splits that characterize the ideology have been fuelled by hatred of the female, since it is through her body that corporeal life continues. Consequently, generations of women, raised in and identifying with the patriarchy, have been unable to provide the loving ground of being for their offspring. Without real choice regarding their own reproductive capacities or their own roles in a society that simultaneously despises and glorifies woman, generations of women have been dissociated from their bodies, ashamed of, and / or penalized for being women.

This situation has had profound effects on the first matrix that each of us experiences: the personal mother. In *Addiction To Perfection*, Jungian analyst Marion Woodman writes:

> Most of our mothers "loved" us and did the very best they could to give us a good foundation for life. Most of their mothers from generation to generation did the same, but the fact remains that most people in this generation, male and female, do not have a strong maternal matrix out of which to go forward into life. Many of our mothers and grandmothers ... longed to be men: some related to their masculine side and dominated the household with masculine values so that the atmosphere was geared to order, to goal-oriented ideals, to success in life, success that they themselves felt they had missed. The gall of their disappointment their children drank with their mother's milk. Unrelated to their own feminine principle, these mothers could not pass on their joy in living, their faith in being, their trust in life as it is.[4]

This is not in any way to blame mothers. It is, rather, to focus on the fact that patriarchal capitalism has, for generations, made it virtually impossible for that first matrix, the personal mother, to be able to provide loving containment and non-poisonous nurturing for her offspring. Having not had it herself, from a mother similarly raised in dissociation from the feminine principle and her own body, the woman cannot lovingly ground her children in their own corporeality. That is supposed to be the fundamental legacy of the first matrix: that the individual feels fully at home in his or her body, loved for this corporeality in all its imperfection and messiness. Without that grounding in the body a person feels uncontained, insecure, continually anxious, and lacking. What is missing is the deep sense of bodily containment, the fundamental assurance that it is acceptable to be in the world, and to be one's unique, bodily self.

Traditionally, the second matrix – human culture – is a development from the first. It, too, encompasses and nurtures the individual, further grounding her or him in larger meaning and social interaction. In traditional societies, and in prepatriarchal matriarchies, people did not feel isolated and separate in the sense of self-alienation and exclusion. Rather, they had a meaningful place within this matrix, which provided containment for individual and social energies, and for their expression through ritual.

Arguably, under patriarchal capitalism, both these necessary matrices have been eroded and distorted: with the result that most people in Western society have been frustrated virtually from birth – unwanted, uncontained, and unnurtured. The lack of grounding in the body leads us to live mainly "in the head", relating to the world through a rational mind split off from bodily wisdom and bodily needs. This splitting-off has an obviously distorting effect on rationality itself. As Hans Magnus Enzensberger writes, "You don't have to be Hegel to catch on to the fact that Reason is both reasonable and against Reason."[5]

What we must recognize is that it is not reason which is to blame for the terrible woes under patriarchy, but rather the splitting-off of reason from the body and human feeling. Obviously, only an ungrounded reason could rationalize the numerous insanities that characterize our way of "life". As Alexander Lowen writes in Depression And The Body: "Knowledge was so important to the development of civilization that it seemed justifiable to deny the body's claim to equality. We are beginning to discover that this was a serious error."[6]

But the loss of the body and the hatred of the feminine have served patriarchal capitalism well. The desperation that results from a lack of grounding leads us to search endlessly for the primary satisfaction and containment that should have come from the first and second matrices as a child. While this is not a new phenomenon, because patriarchy is not new, what *is* new is the omnipresent technological environment which provides a quite unloving substitute "matrix" from which further dissociation and disembodiment result.

□

As the new ritual container, the new "matrix", technological systems offer containment within their global frameworks, but it is a containment that actually isolates and dehumanizes people, who become part of a global mass but are in no real way related to others through the technologies except in the most abstract sense. So, for example, the A.C. Nielsen Company has been deluged with requests from people who want to become part of the company's "single-source data collection" statistical base.[7] Presumably, it is the desire to become part of some "community" that is behind such requests. It is only in the absence of real human community that such substitutes can thrive. Similarly, in *The Second Self*, Sherry Turkle documents the great appeal that computer networks have for adolescent "hackers" who are otherwise lonely, isolated, asocial individuals.[8] Never needing to meet face-to-face for actual interrelationship and communication, computer "freaks" can initiate and terminate "relationships" at the flick of a switch, operating according to the *remote control* that is the zeitgeist of our times.

In the society of the spectacle, the screen is the site of that remoteness, providing both the illusion of "involvement" and the illusion of "control". As our actual impact on decision-making in the corporate, political, and militarist spheres of society has decreased, the illusion of our "control" has been enhanced through an array of interactive and consumer technologies – television, video-games, personal computers, microwave ovens – that seem instantly responsive to our needs, servo-mechanisms for our desires. What we fail to recognize is that in the technological society whose agenda is the expansion of the machine-world into every aspect of life, it is we who are responsive to the "needs" of machines. It is we who are being made into servo-mechanisms of the technologies, each of which perpetuates the mind / body, spirit / matter dichotomies that have been central to the dominant ideology.

By technologizing the domestic sphere and private leisure time with electronics, patriarchal capitalism prepares us to accept the widespread automation and machine-tending that is its ideal for the future of the workplace, production, and consumption. The intrusive computerization of the office through the word-processor, for example, has already mechanized and standardized the work of many women, who are monitored by the machine, marginalized by its digital efficiency, and eventually will be eliminated by its superior functioning.[9] It seems to be women – as secretaries, long-distance operators, telephone switchboard operators, bank tellers – who have been first targeted for replacement by computer. But the advancements in robotics and automation are obviously aimed at most work which in the past has been unionized. The replacement of the human worker by machine neatly resolves corporate problems of long-term economies and control.

We are expected to be complaisant about these developments and not question them. Supposedly they bear the stamp of the inevitable and will usher in an age of abundant materialism. Behind this agenda, however, is the complete marginalization of human beings and all living species: the patriarchal desire to remake the planet into a more efficient and "perfect" simulacrum. As Monica Sjöö and Barbara Mor point out, "Materialism does not glory in matter, but in the male's ability to manipulate it. The new materialism is machine-worship, and product-worship, not the vital and ecstatic celebration of spirit-matter that characterized the Goddess religion" of prepatriarchal times.[10] But the necrophilia of capitalist patriarchy is becoming increasingly difficult to overlook. Marion Woodman writes:

> Our generation is a bridge generation attempting to make a giant stride in consciousness. Faced with atomic power, faced with the possibility of our own self-destruction, we are trying to reconnect to roots that have lain dormant underground for centuries in the hope that the nourishment from those depths may somehow counter-balance the sterility of the perfect machine.[11]

For Woodman and a growing number of feminists, these roots lead to the repressed feminine principle, which patriarchy has successfully obliterated from (male and female) consciousness. In Jungian terms, the individual man or woman is psychologically both masculine and feminine. In the healthy individual, the inner marriage between these two psychic principles results in a self whose actions and thinking are rooted in bodily wisdom and connected awareness of the biosphere. Over the generations, patriarchy has destroyed the balance between these principles, obliterating the feminine principle so that its energies in no way temper the energies of the masculine. As Sjöö and Mor put it: "Patriarchal societies are founded upon a crime. This crime is not the murder of the father, as Freud would have us believe. It is the rape and scorn of the mother."[12]

The elimination of matriarchal cultures by the patriarchy from about 2,000 B.C. onwards was more than a socio-political conquest. It was an attempt to wipe out all memory and knowledge of the source of life as the Earth's mother-womb. Obviously, that crime still reverberates through contemporary society, not just in the countless forms of violence towards women and in the systematic exploitation and despoiling of the earth, but also in the entire anti-people climate that has resulted from patriarchy's desire to remake the world in the form of technological substitutes: simulacra which "prove" the irrelevance of the female and the superiority of the male to refashion life according to a flesh-free, and thereby more "perfect", criterion. That desired technological goal is ultimately a hatred of the earth and incarnation itself: graphically expressed in the postwar priority to soar off ever further from any grounding in earth as the source of all life. The space programs that so fascinate our techno-engineers reflect that "scorn of the mother" which fuels all New Frontiers. Jung stated in 1961: "Sooner or later man will have to return to earth, and to the land from which he comes; that is to say, man will have to return to himself. Space flights are merely an escape, a fleeing away from oneself, because it is easier to go to Mars or to the moon than it is to penetrate one's own being."[13] But obviously, what is being fled from and escaped is the feminine principle, the female side of men themselves.

By obliterating the feminine throughout its history, patriarchy has also twisted and distorted the masculine principle of the psyche – turning it into a cruel, power-mad, disconnected, unfeeling caricature. Given patriarchy's distortions, we see in our own time the extent to which many of us are struggling to know what it really means to be a woman or a man. Our role models have been inadequate, and our media stereotypes have been determined by patriarchal capitalism itself. If, as Woodman says, we are a "bridge generation", the reconnections we must make are in the healing and resolution of false splits and dissociations that have long structured our society.

Nevertheless, given the necrophilia of the technological agenda, it is the feminine principle that immediately needs to be addressed and healed, understood and reconnected with in a conscious way. This is crucial because patriarchal capitalism, having excluded and repressed the feminine and its wisdom, has substituted global technological systems in its place. These systems, arguably, are anima-projections of the male psyche – unconsciously reflecting what the patriarchy considers "feminine". They are nonetheless effective and affective in their twisted guise. Feminists tend to focus on the more obvious penile aspects of technologies: certainly evident in all the missiles and rockets, with their explosive thrusts. What we have overlooked, however, is the extent to which technological systems imitate the "feminine" and are thereby highly appealing in a society that excludes the real feminine principle and makes women themselves peripheral.

☐

Television technology is like the "good whore": readily accessible, instantly "on", inexpensive and non-demanding, providing pleasure and momentary satisfaction, promising more if we stay tuned. From another angle, television is like the "good mother": available twenty-four hours a day, catering to our "needs", comforting, reassuring and entertaining us, even holding and returning our gaze just like in the mother-child dynamic.

As a system, television provides a technological imitation of all-embracing containment. We slip into its soothing, electronic embrace like a foetus floating in the eternal "now" of its electronic ocean, its ever-present "flow" that surrounds us and buoys us up. It seems to contain the whole world and thus it is the new "environment", replacing the natural biosphere which once was the resonant surround within which we lived. Connected to this "source" we are seemingly "nurtured" daily and held in an "intimate" embrace. By providing this womb-like containment, television can thereby "feed" us whatever images and information suit the needs of patriarchal capitalism. As the TV / brain research reveals, we tend to be trusting and unquestioning of this "source", as open to it as a child to its mother. It is in this sense that television is the ultimate frame for the entire technological agenda. It feeds us its bias for the dead object.

Other technological systems similarly imitate the "good mother" as anima-projections. Nuclear power networks provide energy (a function of the womb) and the "security" of being connected to a "source" that is seemingly unlimited and beneficent in its capacity to provide for our needs. Having effectively eliminated the real feminine principle, patriarchy has, throughout this century but especially in the postwar era, patterned its technologies on the imitation of the "feminine", especially the imitation of the "mother". Thus, we are symbiotically connected to, and contained within, immense systems – most of them fittingly brought to us through the efficient umbilical cords of Mother Bell (AT&T): computer networks, cable-TV networks, television networks, people-metering networks, nuclear weapons networks, satellite networks, and (eventually) the "Star Wars" network. All these technological networks provide webs of womb-like containment in a society structured around exclusion of the real feminine principle and the loss of the body.

The foetus's first experience of connectedness and security within a larger, beneficent embrace is provided by the web of uterine tissue which is (in the healthy mother) biologically responsive to, and non-rejecting of, the foetal body. Our technological networks – TV networks are called "webs" in *Variety* – imitate this enmeshing tissue and this non-rejecting response by appearing as "user-friendly" matrices whose nets of containment are secure, enduring, stable. They thereby receive the mass-projections of the mother archetype in a society otherwise devoid of containment and the feminine.

Undoubtedly, this patterning of technological systems to imitate the

"mother" was at first unconscious: the unaware anima-projection of scientists and engineers. But there is now every indication that patriarchal
capitalism is fully conscious of the benefits to be gained from that imitation. The TV / brain research of the late 1960s revealed that the shutting-
down of the left hemisphere allowed access to right hemisphere resonance:
that side of the neo-cortex which is in tune with entire gestalts: the greater
interconnected wholeness that overrides parts, separateness, digital
boundaries. By 1975, scientific research had revealed that technologically
re-establishing early infantile unity with "the mother" yielded interesting
behaviour results. Subjects who received the subliminal message,
"Mommy and I are one", flashed on a tachistoscope screen were able to
overcome previous emotional and behavioural problems.[14] Arguably, in
our society technological systems are the "mommy" with which we are
made one. Televison is purposely used in this way by the industry, exploiting the mass-projections of the archetypal Mother which the technology
triggers.

Rather than actually alter the society to make it more humane, more
balanced, more biologically sane, patriarchy must instead exploit what is
missing. Other technologies are pitched to us on the basis of their "feminine" imitation. So, for example, computer-lingo refers to new "generations" of the machines, new "offspring". Even automobile ads now refer
to new models as "new species". As technology becomes the new "environment", the new "nature", it co-opts characteristics that are properly
non-technological. And, of course, it co-opts the birth process itself, as
Gena Corea has thoroughly documented in her book *The Mother
Machine: Reproductive Technologies From Artificial Insemination to
Artifical Wombs.*[15]

As more and more of life is enclosed within the technological embrace,
we lose a grounded stance from which to differentiate ourselves, much less
curtail and control or eliminate, the encircling growth of these systems.
Their numinosity and popularity depend on the exploitation of the mother
archetype, and perhaps even of the ancient memory of the "Great
Mother", whose religions and societies were actually killed off by the
patriarchy itself. Many (primarily male) enthusiasts of global technologies
like television and the computer speak of the "cosmic oneness" and the
"planetary culture" that will arise from these systems. These enthusiasts
don't seem to recognize that the resulting "oneness" has nothing to do
with interconnectedness with the earth, the biosphere, other people, or
even the actual cosmos. It is, instead, an imitation "oneness", a technological homogenization, a machine-interconnectedness that benefits the
power-drives and greed of multinationals and techno-institutions alone.

Moreover, any politicized person living outside the United States can recognize the term "planetary culture" as being a buzzword for American imperialism.

Thus, the four thousand years-long effort to eradicate the feminine principle continues in its most subtle form: having the technological systems themselves imitate the all-inclusive embrace of the mother. Thus, we have reached what might be seen as an ironic moment within patriarchy itself, were it not for the fact that it is patriarchal capitalism that benefits from the irony.

Over the centuries, the hallmark of patriarchy has been the individual consciousness: a necessary differentiation from the collective embrace and unconscious immersion in the natural world – the embrace that seems to have characterized the species in its earliest ages. The patriarchy has always feared the return of that ancient era, a fear expressed by C.S. Lewis: "In the hive and the ant-hill we see fully realized the two things that some of us most dread for our own species – the dominance of the female and the dominance of the collective."[16] Whether or not that is an accurate description of the species fifty thousand or more years ago, I will leave to feminist archaeologists to debate. Nevertheless, at some point in the species' past, there was a step out of the collective and unconscious embrace of nature – a step necessary for the maturation of the species, just as the child must separate itself from the symbiotic embrace of the mother over a period of years. Patriarchy seized on this separation and the resulting individual consciousness as the fundamental basis for its zeitgeist.

Now, however, as patriarchal capitalism remakes all nature into technological simulacra and fashions its technological systems to imitate the "feminine", its purpose is to sweep us all into an unconscious collectivity exploited by the power drives of the elite few. Individual consciousness – once the very hallmark of the patriarchy – is no longer prized in such a technocracy, which has replaced the feminine principle with the allure of the techno-"feminine". We are forfeiting individual consciousness to a technological totalitarianism that is appealing precisely because of its "feminine" imitation of containment and interconnecting "oneness".

In this guise, patriarchy's techno-systems actually despoil the planet, wipe out indigenous cultures, homogenize the mind, eliminate the body, exploit the unconscious, and create a flesh-free environment in which all living beings are in the way of "progress". Patriarchy no longer fears the hive and the ant-hill as the fate of the species because technology and the techno-elite have become the "queen bee".

At the same time, our patriarchal leaders have now embarked on their goal of cutting back and / or eliminating all programs for social assistance and financial help aimed at those (most of us) who find it difficult to survive in such an "environment". In a revealing phrase, Margaret Thatcher has referred to "the nanny state" – her disparaging description of state-provided social programs which, in her mind, go against so-called self-reliance and individual initiative.[17] Although Thatcher can recognize the patriarchal state's imitation of the "feminine" in such assistance programs – and actually uses that recognition as the basis for belittlement – she would undoubtedly never recognize or admit that the industries and global techno-systems she supports are themselves patterned after a similar imitation of the "feminine" – receiving and exploiting the mass-projection of the mother archetype in a society utterly devoid of the real feminine principle. It must also be stated that if there is any group which has been financially dependent for handouts from "the nanny state", it has certainly been the nuclear industry – which for forty years has received massive assistance from our militarist and technophilic governments.

□

Those of us who recognize the horrible absurdities that characterize patriarchal capitalism and its notion of "progress" are going to have to take a stand with regard to the technological agenda that now determines every facet of existence. As Hannah Arendt reminds us, "The aim of totalitarian education has never been to instill convictions but to destroy the capacity to form any."[18] By 1986 our world leaders were spending $800 billion per year on the arms race, or $1.3 million every minute.[19] That expenditure is justified by the "fact" of hostile superpowers divvying up the world and threatening one another's sacrosanct ideologies.

At the level of technological systems we might more accurately see the situation from another angle. A case in point is the global nuclear fuel cycle. The process of turning raw uranium into the fissionable material necessary for bombs and nuclear power plants is an international process that involves "as many as five countries at a time in the production of a single fuel element".[20] The uranium must go from the mines to refineries to enrichment factories to conversion facilities to fabrication plants to reactors to reprocessing plants to weapons manufacturing facilities. The most costly stage is the enrichment process and there is a competitive world market in bidding for contracts to enrich Western uranium.

By 1976, the Soviet Union had entered the bidding and began to land contracts because its facilities were able to do the enriching more cheaply. At the same time, other stages in this "furtive underground global network" of uranium processing have been handled by NATO countries, with the result that U.S. missiles aimed at Russia may be fuelled by fissionable material processed by the Soviets, and Soviet missiles aimed at the United

States likely contain nuclear material mined and partly processed in North America.[21] As Bob Hunter, author of "Plutonium in Motion", writes:

> In the Byzantine world of the nuclear fuel industry it is not surprising that even as the Soviet Army plunged into Afghanistan and Jimmy Carter cancelled shipments of grain to the U.S.S.R., some two-and-a-half tons of UF6, fresh from an old Russian nuclear weapons plant, was being off-loaded from a French steamer in Seattle for shipment to the Exxon nuclear facility at Richland, near Hanford, Washington.[22]

That Byzantine world has not been altered by any change in government leadership, since it operates at the level of multinational corporate co-ordination and according to the economies of technological efficiency. There, a different ideology rules – one that has little to do with supposedly hostile superpowers. Rather, the hostility is directed at us – the millions of human beings across the planet whose corporeal needs are in the way of technological expansion and whose human emotions are an embarrassment to the global techno-institutions, unless those emotions can be usefully manipulated.

What we are up against is the mind-set epitomized on an evening in the summer of 1945, the evening before the first test of the atomic bomb. The nuclear scientists themselves had no idea whether or not the atomic blast would actually ignite the earth's atmosphere and thus destroy the entire planet. Nevertheless, they were quite willing to go ahead with the test and find out. The evening before the atomic blast at dawn, the brilliant physicist Enrico Fermi offered to take bets on the outcome.[23] No sacrifice is too great for the construction of the perfect machine.

Notes

Chapter 1 Introduction: The Bolt in the Soul

1 Edward Zuckerman, *The Day After World War III* (New York: Avon Books, 1987), p. 100. Zuckerman's extraordinary text is a fully documented portrait of the FEMA contingency plans for "business as usual" in a post-nuclear USA.

2 Quoted in Leslie J. Freeman (ed.), *Nuclear Witnesses* (New York / London: W.W. Norton & Company, 1982), p. 48.

3 Fritjof Capra, *The Turning Point* (New York: Bantam, 1982), p. 242.

4 Quoted in Freeman, *Nuclear Witnesses*, p. 69.

5 Kevin Robins and Frank Webster, "Luddism: New Technology and the Critique of Political Economy," in Les Levidow and Bob Young (eds.), *Science, Technology and the Labour Process*, vol. 2 (London: Free Association Books, 1985), p. 28.

6 Jeremy Rifkin, *Declaration of a Heretic* (Boston: Routledge & Kegan Paul, 1985), p. 23.

7 Ibid., p. 66.

8 George Grant, "The Morals of Modern Technology," *The Canadian Forum* (October 1986), p. 17.

9 Francis Barker, *The Tremulous Private Body* (London: Methuen, 1984), p. 67.

10 Marion Woodman, *Addiction to Perfection* (Toronto: Inner City Books, 1982), p. 16.

11 Dr. Ursula Franklin, "The Second Scientist," *The Canadian Forum* (December 1985), p. 53.

12 Mary Shelley, *Frankenstein* (Oxford: Oxford University Press, 1969), pp. 47-48.

Emphasis mine. Brian Easlea, *Fathering the Unthinkable: Masculinity, Scientists and the Nuclear Arms Race* (London: Pluto Press, 1983) also uses Shelley's text as the basis of a critique of masculine science, but from a slightly different angle: that of men imitating women's reproductive role.

13 Ibid., p. 49. Emphasis mine.

14 Ibid., p. 51.

15 Ibid., p. 54.

16 Ibid., p. 52.

17 Ibid., p. 51.

18 Ibid., p. 55.

19 Ibid., p. 54.

20 Ibid., p. 160.

21 Quoted in Capra, *Turning Point*, p. 59.

22 Quoted in Elizabeth Dodson Gray, *Green Paradise Lost* (Wellesley, Mass.: Roundtable Press, 1979), pp. 22-23.

23 Ibid.

24 Arthur Kroker, *Technology and the Canadian Mind: Innis / McLuhan / Grant* (Montreal: New World Press, 1984), p. 127.

25 Ibid., p. 128.

26 Edward C. Whitmont, *The Symbolic Quest* (Princeton: Princeton University Press, 1978), pp. 28-29. Emphasis mine.

27 Quoted in Kroker, *Technology*, p. 67.

Chapter 2 The New World of the Bomb

1 I.F. Stone, *The Haunted Fifties* (New York: Vintage Books, 1969), p. 120.

2 Quoted in Connie Tadros, "Film, TV and the Nuclear Apocalypse: Peter Watkins on Media," *Cinema Canada* (October 1984), p. 22

3 Stephen Hilgartner, Richard C. Bell, Rory O'Connor, *Nukespeak: The Selling of Nuclear Technology in America* (New York: Penguin Books, 1983), p. 22.

4 Ibid., p. 25.

5 Ibid., p. 26.

6 Vincent Leo, "The Mushroom Cloud Photograph: From Fact to Symbol," *AfterImage* (Summer 1985), pp. 6-12; and Wilfred Burchett, *Shadows of Hiroshima* (London: Verso Editions, 1983).

7 Leo, "Mushroom Cloud," p. 7.

8 Ibid., pp. 7-8.

9 Ibid., pp. 7, 8.

10 Rosalie Bertell, *No Immediate Danger?* (Toronto: Women's Educational Press, 1985), p. 138. See also Peter Wyden, *Day One: Before Hiroshima and After* (New York: Warner Books, 1985), p. 327; Wyden notes that documentary footage shot by Japanese producer Akira Iwasaki, some 45,000 feet of film, was also confiscated by the U.S. military and classified as "top secret".

11 Ibid., pp. 150-151.

12 Erik Barnouw, *Tube of Plenty* (New York: Oxford University Press, 1975), p. 85.
13 Quoted in Peter Pringle and James Spigelman, *The Nuclear Barons* (London: Sphere Books, 1983), p. 8.
14 Hilgartner, Bell, O'Connor, *Nukespeak*, p. 72. Emphasis mine.
15 Quoted in Ibid., pp. 72-73.
16 Bertell, *No Immediate Danger?*, p. 74.
17 Ibid., p. 137.
18 Easlea, *Fathering the Unthinkable*, p. 120.
19 Quoted in Paul Boyer, *By the Bomb's Early Light* (New York: Pantheon Books, 1985), p. 14.
20 Ibid.
21 Barnouw, *Tube of Plenty*, p. 108.
22 Ibid., p. 112.
23 Ibid., pp. 180-181.
24 Quoted in Barbara Ehrenreich, *The Hearts of Men* (New York: Anchor / Doubleday, 1983), p. 45.
25 Stuart Ewen, *Captains of Consciousness* (New York: McGraw-Hill, 1976), p. 213. Emphasis mine.
26 Marshall McLuhan, *Understanding Media* (Toronto: McGraw-Hill, 1965), p. 41.

Chapter 3   Atomic Fictions

1 Thomas H. Saffer and Orville E. Kelly, *Countdown Zero* (Harmondsworth: Penguin Books, 1982), pp. 43-44. See also Howard L. Rosenberg, *Atomic Soldiers* (Boston: Beacon Press, 1980).
2 Erik Barnouw, *The Sponsor* (New York: Oxford University Press, 1978), p. 162. Barnouw was instrumental in bringing to American attention the confiscated footage shot by Japanese producer Akira Iwasaki in Hiroshima and Nagasaki. He devotes a section of this text to the U.S. TV networks' co-operation with the Atomic Energy Commission through the 1950s and 1960s (pp. 149-170).
3 Gregory Bateson, *Steps to an Ecology of Mind* (New York: Ballantine Books, 1972).
4 Quoted in Freeman, *Nuclear Witnesses*, p. 294.
5 Barnouw, *The Sponsor*, p. 163.
6 Ibid.
7 Hilgartner, Bell, O'Connor, *Nukespeak*, p. 44.
8 Bertell, *No Immediate Danger?*, p. 171.
9 Barnouw, *The Sponsor*, p. 45.
10 Barnouw, *Tube of Plenty*, p. 160.
11 Erik Barnouw, interview with the author, Washington, D.C., 1981.
12 Ibid.
13 Quoted in William Boddy, "Entering *The Twilight Zone*," *Screen* (July-October 1985), p. 100.

14 Barnouw, *The Sponsor*, pp. 108-109.
15 David Susskind, interview in CBC archives, Toronto, 1960.
16 Noam Chomsky, "The Bounds of Thinkable Thought," *The Progressive* (October 1985), pp. 28-31.
17 John Leonard, interview with the author, New York, 1977.
18 Norman Campbell, interview with the author, Toronto, 1977.
19 Bob Shanks, "Network Television: Advertising Agencies and Sponsors," in John Wright (ed.), *The Commercial Connection* (New York: Dell / Delta, 1979), p. 94.

Chapter 4   The Sweetening Machine

1 Ron Rosenbaum, "Kanned Laffter," in James Monaco (ed.), *Media Culture* (New York: Delta Books, 1978), p. 133.
2 Peter Campbell, interview with the author, Toronto, 1984.
3 Joe Partington, interview with the author, Toronto, 1984.
4 Rosenbaum, "Kanned Laffter," p. 140.
5 Guy Debord, *Society of the Spectacle* (Detroit: Black and Red, 1983), p. 20.

Chapter 5   Fine-Tuning: TV, Desire, and the Brain

1 Tony Schwartz, *The Responsive Chord* (Garden City, N.Y.: Anchor Books, 1973).
2 Quoted in Ed Magnuson, "Taking Those Spot Shots," *Time* (September 29, 1980), p. 31.
3 Ibid.
4 Quoted in Michael J. Arlen, "The Air," *The New Yorker* (February 18, 1980), p. 108.
5 Schwartz, *Responsive Chord*, p. 17.
6 Herbert E. Krugman, *Electroencephalographic Aspects Of Low Involvement: Implications For The McLuhan Hypothesis* (New York: American Association for Public Opinion Research, 1970).
7 Ibid., p. 17.
8 Tony Schwartz, interview with the author, New York, 1977.
9 Fred Emery and Merrelyn Emery, *A Choice of Futures* (Canberra: Australian National University, 1975).
10 Robert E. Ornstein, *The Psychology Of Consciousness* (San Francisco: W.H. Freeman & Company, 1972).
11 See especially *Understanding Media*; and *Counter Blast* (Toronto: McClelland and Stewart, 1969).
12 Schwartz, *Responsive Chord*, p. 69.
13 Marion Woodman, *The Pregnant Virgin* (Toronto: Inner City Books, 1985), p. 20.
14 Tony Schwartz, *Media: The Second God* (Garden City, N.Y.: Anchor Books,

1983); see also the Schwartz quotation in the chapter entitled "The Righteous Stuff: Evangelism and the TV God".

15 Julian Jaynes, *The Origin Of Consciousness In The Breakdown Of The Bicameral Mind* (Toronto: University of Toronto Press, 1976), pp. 202-203.

16 Quoted in Michael Valpy, "Beyond the Image," *The Globe and Mail* (March 5, 1983), p. 5.

17 Quoted in Roy MacGregor, "The Selling Of The Candidates," *Maclean's* (May 7, 1979), p. 28.

18 Quoted in Magnuson, "Taking Those Spot Shots," p. 33.

19 William A. Henry III, "Mirror, Mirror On The Tube," *Time* (August 17, 1981), p. 42.

20 Arlen, "The Air," p. 112.

21 Quoted in David Chagall, *The New Kingmakers* (New York: Harcourt Brace Jovanovich, 1981), p. 301.

22 Quoted in Henry III, "Mirror, Mirror," p. 43.

23 Schwartz, *The Responsive Chord*, pp. 63, 65.

Chapter 6   Watching For The Magoos

1 Earl C. Gottschalk Jr., "Testing Of TV Shows And Movies Is Gaining Adherents and Critics," *The Wall Street Journal* (January 12, 1981).

2 Roland Flamini, "Television and the Magoo Factor," *Journal of American Film* (May 1976), p. 62.

3 Ibid., p. 63.

4 Ibid.

5 Neil Shister, "Polygraphs and Wires: TV Research Guys Mean Business," *More* (December 1976), p. 39.

6 Harry F. Walters, "How Research Tyrannizes TV," *Newsweek* (August 29, 1983), p. 53.

7 Shister, "Polygraphs and Wires," p. 40.

8 Walters, "How Research Tyrannizes TV," p. 53.

Chapter 7   The Righteous Stuff: Evangelism and the TV God

1 Quoted in John Picton, "Prayer Over the Air: A $1 Billion Business," *The Toronto Star* (February 21, 1982).

2 Quoted in Johnny Greene, "The Astonishing Wrongs of the New Moral Right," *Playboy* (January 1981), p. 118.

3 Quoted in Joe Hall, "Righteous TV Preacher Grabs Power With Reagan," *The Toronto Star* (November 15, 1980).

4 Quoted in Greene, "Astonishing Wrongs," p. 252.

5 Quoted in Val Sears, "The Armies Of The Right," *The Toronto Star* (December 1, 1981).

6 "Interview: Reverend Jerry Falwell," *Penthouse* (March 1981), p. 152.
7 Quoted in Greene, "Astonishing Wrongs," p. 252.
8 "Interview," *Penthouse,* p. 154.
9 Quoted in Richard N. Ostling, "Power, Glory – And Politics," *Time* (February 17, 1986), p. 55.
10 Quoted in Greene, "Astonishing Wrongs," p. 118.
11 Quoted in Ostling, "Power, Glory," p. 58.
12 Quoted in Greene, "Astonishing Wrongs," p. 118.
13 Quoted in Joe Klein, "The Moral Majority's Man In New York," *New York* (May 18, 1981), p. 30.
14 Quoted in Ostling, "Power, Glory," p. 58.
15 Quoted in Sara Diamond, "Religious Broadcasters Broaden Their Sights," Cooperative News Service Release (February 1986).
16 Quoted in Ostling, "Power, Glory," p. 58.
17 Quoted in Michael D'Antonio, "Christian America Is On The March," *The Toronto Star* (January 18, 1986).
18 Quoted in Neil Louitt and Tim Harpur, "Moral Rebirth Due – Falwell," *The Toronto Star* (October 25, 1982).
19 Quoted in Erik Knutzen, "Prime-Time Proselytizers," *Starweek* (December 20, 1980), p. 11.
20 Quoted in Albert Moritz, "The Lord Is My Channel Commander," *Weekend Magazine* insert in *The Toronto Star*; my apologies to Albert Moritz: I am unable to locate the date for this clipping.
21 Quoted in Margaret O'Brien Steinfels and Peter Steinfels, "The New Awakening: Getting Religion In The Video Age," *Channels* (January-February 1983), p. 26.
22 Quoted in Knutzen, "Prime-Time Proselytizers," p. 9.
23 Quoted in Ostling, "Power, Glory," p. 55.
24 Quoted in Val Sears, "TV Evangelists Exposed," *The Toronto Star* (August 30, 1981).
25 Quoted in Schwartz, *Media,* p. 4.
26 Ibid., p. 1.
27 Quoted in Moritz, "The Lord," p. 12.
28 Steinfels and Steinfels, "New Awakening," p. 26.
29 "Interview," *Penthouse,* p. 150.
30 Quoted in Jeffrey Herf, *Reactionary Modernism: Technology, Culture, and Politics in Weimar and The Third Reich* (New York: Cambridge University Press, 1984), p. 196. Herf's emphasis.
31 Quoted in Hall, "Righteous TV Preacher."
32 Alan Crawford, *Thunder On The Right* (New York: Pantheon Books, 1980).
33 Quoted in "David Frost To Reappear With The Next President," New York Times Service feature published in *The Globe and Mail* (May 25, 1987).

Chapter 8   TV News: A Structure Of Reassurance

1 Wallace Westfield, interview with the author, New York, 1977.
2 Edward R. Murrow, *See It Now* (November 18, 1951), CBS Program Archives.
3 Todd Gitlin, *The Whole World Is Watching* (Berkeley: University of California Press, 1980), p. 99.
4 Michael Arlen, interview with the author, New York, 1977.
5 Ibid.
6 Arthur Asa Berger, interview with the author, San Francisco, 1977.
7 Quoted in Edwin Diamond, *Sign Off* (Cambridge: MIT Press, 1982), pp. 70-71.
8 Ibid., p. 18.

Chapter 9   An Electronic Marriage

1 Les Brown, "Buying and Selling the TV Viewer," *Harper's* (January 1986), p. 70.
2 Milton Moskowitz, Michael Katz, Robert Levering (eds.), *Everybody's Business* (San Francisco: Harper and Row, 1980), p. 353.
3 Frank Mankiewitcz and Joel Swerdlow, "Ratings," in John W. Wright (ed.), *The Commercial Connection* (New York: Delta Books, 1979), p. 114. David Chagall, "Can You Believe the Ratings?" in *The Commercial Connection*, p. 218.
4 Brown, p. 71.
5 Quoted in "A.C. Nielsen Plans To Test New Concept," *The Globe and Mail* (April 10, 1985).
6 Wendy Miles, interview with the author, Toronto, 1985.
7 Quoted in David Chagall, *The New Kingmakers* (New York: Harcourt Brace Jovanovich, 1981), p. 184.
8 Will Ellsworth-Jones and Roland Perry, "The Man Who Programs The President," *The Toronto Star* (May 6, 1984); for a more detailed view, see Roland Perry, *Hidden Power* (New York: Beaufort Books, 1984).
9 Ibid.
10 Quoted in Ibid.
11 Lynne Olson, 'The Reticence of Ronald Reagan," *Washington Journalism Review* (November 1981), p. 42.
12 Michael Posner, "Playing the Presidency By Cue," *Maclean's* (November 24, 1981), p. 33.
13 Schwartz, *The Responsive Chord*, p. 82.

Chapter 10   The Global Pillage

1 Canada has been placed in the paradoxical position of being considered by the U.S. industry as part of its domestic market in terms of political lobbying and branch-plant production by the U.S. studios; but Canada is considered "foreign" when it comes to sales of TV shows during their first run on U.S. networks.
2 Barnouw, *The Sponsor*, p. 161.
3 Ibid., pp. 173-174.

4 Fred H. Knelman, *Nuclear Energy: The Unforgiving Technology* (Edmonton: Hurtig Publishers, 1976), p. 74.

5 Andrew R. Horowitz, "The Global Bonanza of American TV," in Monaco, *Media Culture*, p. 119.

6 Alan Wells, *Picture-Tube Imperialism?* (Maryknoll, N.Y.: Orbis Books, 1972), p. 73.

7 Sarnoff produced a pamphlet, *Political Offensive Against World Communism*, cited in Milton Moskowitz, Michael Katz, Robert Levering (eds.), *Everybody's Business* (San Francisco: Harper and Row, 1980), p. 844.

8 Horowitz, "The Global Bonanza," p. 121.

9 Barnouw, *Tube Of Plenty*, p. 230.

10 Horowitz, "The Global Bonanza," p. 121.

11 Ibid.

12 Wells, *Picture-Tube Imperialism?*, p. 179.

13 Horowitz, "The Global Bonanza," p. 121.

14 Brian Murphy, *The World Wired Up* (London: Comedia, 1983), p. 21.

15 Quoted in Harry Skornia and Jack Kitson (eds.), *Problems And Controversies In TV And Radio* (Palo Alto: Pacific Books, 1968).

16 John Telbel, "U.S. TV Abroad: Big New Business," *Saturday Review* (July 14, 1962), p. 33.

17 See especially Peter Morris, "Electronic Free Trade: How The CBC Brought U.S. Television To Canada," *Cinema Canada* (December 1986); and Joyce Nelson, "CRTC Asleep At The Wheel: Canadian Broadcasting In The Age Of Television Sprawl," *Fuse* (Winter 1986 / 87).

18 Kaarle Nordenstreng and Tapio Varis, *Television-Traffic: A One-Way Street?* (New York / Paris: UNESCO, 1974).

19 William H. Read, *Journal Of Communications* (Summer 1976).

20 Jeremy Tunstall, *The Media Are American* (London: Constable Press, 1977), p. 130.

21 Ben Bagdikian, *The Media Monopoly* (Boston: Beacon Press, 1983), p. 20.

22 Horowitz, "The Global Bonanza," p. 118.

23 *Variety* (July 12, 1978).

24 *Variety* (June 21, 1978).

25 Joyce Nelson, "Losing It In The Lobby: Entertainment And Free Trade," *This Magazine* (November 1986).

26 Hilgartner, Bell, O'Connor, *Nukespeak*, pp. 74-75.

27 Quoted in Jamake Highwater, *The Primal Mind* (New York: Meridian Books, 1981), p. i.

Chapter 11 Nuclear TV

1 Quoted in Freeman, *Nuclear Witnesses*, p. 22.

2 Quoted in Hilgartner, Bell, O'Connor, *Nukespeak*, p. 90.

3 "x-Rays in the Living Room," *Time* (August 4, 1967), p. 62.
4 Quoted in Ben Funk, "The Battle Against TV Radiation – It's Just Begun," Associated Press Story (April 24, 1970), published in John Ott, *Health And Light* (New York: Pocket Books, 1976), p. 136.
5 Quoted in Ibid.
6 J.G. Ello, "Measuring Color-TV Generated x-Rays," *Electronics World* (July 1971), p. 37.
7 "x-Rays in the Living Room," p. 62.
8 Don Ward, "TV x-Rays," *Radio-Electronics* (April 1970), p. 55.
9 Ibid.
10 Ibid.
11 Dr. Jan Paul, "Is Your TV A Radiation Hazard?" *Popular Mechanics* (January 1962), p. 201.
12 Ott, *Health And Light*, p. 125.
13 Ibid., p. 126.
14 Ibid., p. 127.
15 Ibid.
16 Quoted in Ibid., p. 130.
17 Quoted in Freeman, *Nuclear Witnesses*, p. 31.
18 Quoted in Ott, *Health and Light*, p. 131.
19 Bertell, *No Immediate Danger?*, p. 224.
20 Quoted in "Perils of TV," *Newsweek* (August 7, 1967), p. 53.
21 Gene Smith, "General Electric Will Modify 90,000 Large Color TV Sets Against Possible x-Radiation," *The New York Times* (May 19, 1967), p. 17.
22 Quoted in Ibid.
23 "Health Service Advice," *The New York Times* (May 19, 1967), p. 17.
24 "TV That Can Bite," *Business Week* (May 27, 1967), p. 61.
25 "TV Radiation Assessed," *Science News* (July 1, 1967), p. 11.
26 *Consumer Reports* (July 1967), p. 349.
27 "Perils of TV," p. 53.
28 Quoted in "Radiation From TV Sets," *Consumer Reports* (September 1968).
29 Quoted in Funk, "The Battle Against TV Radiation."
30 Quoted in "Radiation From TV Sets."
31 Ibid.
32 "TV Radiation," *Consumer Reports* (February 1969), p. 74.
33 "What Are the Radiation Dangers In Color TV?," *Good Housekeeping* (October 1969), p. 217.
34 "Standards For TV Sets," *Science News* (July 19, 1969).
35 "Some Observations On x-Radiation," *Consumer Reports* (January 1970).
36 Quoted in Funk, "The Battle Against TV Radiation."
37 Quoted in Ibid.
38 Jan. S. Paul, "Radiation From Color TV Receivers, Where Lies The Danger?,"

Consumer Bulletin (December 1970), p. 24.

39 Ott, Health and Light, p. 200; Jerry Mander, Four Arguments For The Elimination Of Television (New York: William Morrow and Company, 1978), pp. 172-191.

40 Bob DeMatteo, Terminal Shock (Toronto: NC Press, 1985), p. 48.

Chapter 12 The Power of Ghosts

1 See, for example, Marshall McLuhan, "What Television is Doing to Us – and Why," The Washington Post (May 15, 1977), p. H5; McLuhan's earlier work up through the 1960s has been well summarized in Kroker, Technology and the Canadian Mind, pp. 52-86.

2 Quoted in Steinfels and Steinfels, "New Awakening," p. 27; Cox's view is used by the Steinfels as a contrast to that of the televangelists.

3 Adrienne Rich, On Lies, Secrets, and Silence (London / New York: W.W. Norton, 1979), pp. 12-13.

4 Scott MacDonald, "The Nuclear War Film: Peter Watkins Interviewed," The Independent (October 1984), p. 24.

5 To my knowledge, the major texts are: André Bazin, "The Ontology of the Photographic Image," in André Bazin, What is Cinema? (Berkeley: University of California Press, 1967); Walter Benjamin, "The Work of Art in the Age of Mechanical Reproduction," in Walter Benjamin, Illuminations (New York: Schocken Books, 1969); Roland Barthes, Camera Lucida (New York: Hill and Wang, 1985); and sections in Steve Neale, Cinema and Technology (Bloomington: Indiana University Press, 1985).

6 Mander, Four Arguments.

7 Jerry Mander, interview conducted by Jill Eisen for the author, Los Angeles, 1982.

8 Since many contemporary problems (radiation, toxic rain, PCBs) are not visibly detectable, we can't really see their damage until there is a corpse of some kind to reveal it. This suggests that a society organized around image-information will continually lag behind the destruction underway.

9 Mander, Four Arguments, p. 287.

10 Todd Gitlin, "The Greatest Story Never Told," Mother Jones (June / July 1987), p. 29.

11 Benjamin, Illuminations, p. 231.

12 Bazin, What is Cinema?, p. 14.

13 Ibid., p. 13.

14 Ibid.

15 Quoted in Robins and Webster, "Luddism," p. 25.

16 Quoted in Neale, Cinema and Technology, p. 38.

17 Ibid., pp. 54-55.

18 Easlea, Fathering the Unthinkable, p. 39.

19 McLuhan, "What Television is Doing," p. H6.
20 Monica Sjöö and Barbara Mor, *The Great Cosmic Mother: Rediscovering the Religion of the Earth* (San Francisco: Harper and Row, 1987), p. 105.
21 Barthes, *Camera Lucida*, p. 92.
22 Marshall McLuhan, *Counter Blast*, p. 30.
23 I am indebted to Michael Dorland for this insight on instant replay.
24 Benjamin, *Illuminations*, p. 241.
25 Ibid., p. 242.
26 For the recognition of this incongruity, I am indebted to Susan Boyd-Bowman, "*The Day After*: Representations of the Nuclear Holocaust," *Screen* (July-October 1984), p. 82.
27 For an interesting spin-off from Maritain's metaphor, see B.W. Powe, "Electric Politics: Pierre Elliott Trudeau in the Age of Angelism," *The Canadian Forum* (May 1987), pp. 8-16.
28 Jacques Derrida, speaking in the film *Ghost Dance*, as cited in *Festival of Festivals Catalogue* (Toronto, 1984).
29 Pringle and Spigelman, *Nuclear Barons*, p. 180.

Chapter 13 The Astronautical Body

1 Jack Miller, "Space: Still Soaring After 25 Years," *The Toronto Star* (April 12, 1986), p. B6.
2 Terence Dickinson, "The Next Threshold," *Equinox* 19 (January-February 1985), p. 54.
3 Ibid., p. 57.
4 David Osborne, "Business In Space," *The Atlantic Monthly* (May 1985), p. 45.
5 The IMAX cinema technology is a Canadian invention. As such, it seems to appropriately signify the one-hundred-year-long fascination of Canadian spectators with the screen. Since that screen has always been the space for colonization, it appears fitting that the Canadian IMAX screen be devoted to the colonization of space.

Chapter 14 Conclusion: The Whirlwind, The Storm

1 Terri Nash says that the French-Canadian network, Radio-Canada, showed the film months before English-CBC and publicized it well. "They didn't need Hollywood's seal of approval" before broadcasting it.
2 This aspect of critical reaction to the production is cited in Boyd-Bowman, "*The Day After*," p. 95.
3 In his film *The Journey*, Peter Watkins notes that the U.S., immediately after the war, established the Atomic Bomb Casualty Commission to study the effects of the blasts, "especially the radiation of victims in Hiroshima". The Commission, however, refused to provide medical treatment for the victims it was studying.
4 Woodman, *Addiction to Perfection*, pp. 16-17.

5 Quoted in Robins and Webster, "Luddism," p. 9.

6 Alexander Lowen, *Depression and the Body* (Harmondsmith: Penguin, 1972), p. 48.

7 Personal communication to the author, 1985.

8 Sherry Turkle, *The Second Self* (New York: Simon and Schuster, 1984).

9 For an expansion on this view, see Heather Menzies, "In His Image: Science and Technology as Ideology," *This Magazine* (May / June 1987), pp. 31-34.

10 Sjöö and Mor, *Great Cosmic Mother*, p. 233.

11 Woodman, *Addiction to Perfection*, p. 17.

12 Sjöö and Mor, *Great Cosmic Mother*, p. 193.

13 William McGuire and R.F.C. Hull (eds.), *C.G. Jung Speaking* (Princeton: Princeton University Press, 1977), p. 468.

14 R.C. Morse and David Stoller, "The Hidden Message That Breaks Habits," *Science Digest* (September 1982), p. 28.

15 Gena Corea, *The Mother Machine* (New York: Harper and Row, 1985).

16 Quoted in William Irwin Thompson, *The Time Falling Bodies Take To Light* (New York: St. Martin's Press, 1981), p. 156.

17 Thatcher was quoted in Richard Gwyn, "The Iron Lady Shows Her Mettle In Historic Win," *The Toronto Star* (June 12, 1987), p. A16.

18 Hannah Arendt, *Totalitarianism* (New York: Harcourt Brace Jovanovich, 1968), p. 166.

19 Peter Watkins, *The Journey*, film transcript, p. 88.

20 Bob Hunter, "Plutonium In Motion," *This Magazine* (November 1984), p. 29.

21 Ibid., p. 31.

22 Ibid., p. 29.

23 Zuckerman, *The Day After World War III*, p. 15.

Other Books from

IMAGES IN ACTION
A Guide to Using Women's Film and Video
Ferne Cristall and Barbara Emanuel

JUMP CUT
Hollywood, Politics and Counter-Cinema
Edited by Peter Steven

ROOTS OF PEACE
The Movement Against Militarism in Canada
Edited by Eric Shragge, Ronald Babin, and Jean-Guy Vaillancourt

SILVER THREADS
Critical Reflections on Growing Old
Doris Marshall

For a complete catalogue write to:
Between The Lines
229 College Street
Toronto, Ontario
Canada M5T 1R4

Printed in Canada